CAUCHY AND THE CREATION OF COMPLEX FUNCTION THEORY

Between 1814 and 1831, the great French mathematician A. L. Cauchy created practically single-handedly a new branch of pure mathematics. Complex function theory was and remains of central importance, and its creation marked the start of one of the most exciting periods in the development of mathematics.

In this book Dr Smithies analyses the process whereby Cauchy created the basic structure of complex analysis, describing first the 18th century background before proceeding to examine the stages of Cauchy's own work, culminating in the proof of the residue theorem and his work on expansions in power series. Smithies describes how Cauchy overcame difficulties including false starts and contradictions brought about by over-ambitious assumptions, as well as the improvements that came about as the subject developed in Cauchy's hands. Controversies associated with the birth of complex function theory are described in detail. Throughout, new light is thrown on Cauchy's thinking during this watershed period.

This book is the first to make use of the whole spectrum of available original sources; it will be recognised as the authoritative work on the creation of complex function theory.

Frank Smithies, a Fellow of St John's College, Cambridge, taught and researched in mathematics at the University for over 40 years, specialising in the area of functional analysis. During that period he has published many papers in research journals, written a book on Integral Equations, and edited a volume of G. H. Hardy's Collected Papers. Over the last 20 years he has been increasingly interested in the history of mathematics, giving numerous lectures on Cauchy. His activities have culminated in the present book.

CAUCHY AND THE CREATION OF COMPLEX FUNCTION THEORY

FRANK SMITHIES

St John's College, Cambridge

CAMBRIDGE
UNIVERSITY PRESS

PUBLISHED BY THE PRESS SYNDICATE OF THE UNIVERSITY OF CAMBRIDGE
The Pitt Building, Trumpington Street, Cambridge CB2 1RP, United Kingdom

CAMBRIDGE UNIVERSITY PRESS
The Edinburgh Building, Cambridge CB2 2RU, United Kingdom
40 West 20th Street, New York, NY 10011-4211, USA
10 Stamford Road, Oakleigh, Melbourne 3166, Australia

First published 1997

Printed in the United Kingdom at the University Press, Cambridge

Typeset in 3B2 in Times 10/13 pt

A catalogue record for this book is available from the British Library

ISBN 0 521 59278 X hardback

#36430598

To the memory of Sir Edmund Whittaker, who first aroused my
interest in the history of mathematics.

Contents

Introduction

I.1. During the years 1814–1831 Cauchy created the basic framework of complex function theory. The aim of the present work is to trace the details of this process, beginning with his 1814 memoir on definite integrals, and taking the story as far as the two Turin memoirs of 1831. That year marks a major break in the development of his work in this direction: it is true that for some time thereafter he was making important applications of his results, but he made no further substantial contributions to the central core of the theory until after his return to teaching in 1849.

The available evidence for Cauchy's ideas lies in his published work, practically all his manuscripts and correspondence having disappeared. I hope to show that a connected story emerges from a careful study of his publications, and that we can trace a gradual development in Cauchy's ideas about complex function theory through a number of stages of his work.

We shall concentrate our attention on the main line of development, and we shall leave a number of topics undiscussed, including, for instance, his work on Fourier transforms and differential equations, some of which does involve applications of his results on complex function theory. Important as these achievements were in their own right, they do not seem to have had any significant influence on his development of the central framework of the theory.

Again, we shall not be discussing Gauss's well-known 1811 letter to Bessel, from which it is clear that he was aware of many of the central facts of complex function theory; since he published nothing about this at the time, the ideas he expressed there can have had no influence on what Cauchy was doing. On the other hand, we shall have something to say about certain ideas concerning functions of a complex variable that were already known before Cauchy's time, and which can be found in

1

the publications of such authors as d'Alembert, Euler, Laplace and Poisson.

Cauchy's papers on complex function theory were published in many different places and did not appear in any kind of logical order; the same result often occurs several times, frequently in a somewhat modified form. He never published any connected exposition of his results in this field.

This confusion is reflected in the secondary literature; most writers concentrate their attention on a small selection of his publications, especially the 1814 memoir (which was not published until 1827), the 1825 memoir on definite integrals between imaginary limits, and the 1841 republication in his *Exercices d'analyse* of the first Turin memoir of 1831, and they may also mention a few of the notes appearing in his *Exercices de mathématiques* in 1826–1830. As a result, they provide no clear chronology of his work, being often inconsistent with one another, especially about the dates to be attributed to particular results. For instance, both Osgood and Felix Klein date Cauchy's theorem on the integral of a function round a closed curve to 1825 (which is too early, and refers to a different result), whereas Brill and Noether give 1846 (which is far too late); as we shall see, 1831 would be a more accurate date. Again, Brill and Noether date the residue theorem for a circle to 1841, although Cauchy stated it explicitly in 1827. Part of the trouble arises from the fact that some of Cauchy's most important results appeared only in his second Turin memoir of 1831, and in brief abstracts in the same year in the *Bulletin de Férussac*; these publications were not widely available.

The situation has greatly improved with the completion of Cauchy's collected works in 1974, and the publication of Belhoste's detailed biography in 1991, making it possible to date Cauchy's results with some precision.

I.2. The present work falls into seven chapters. In the first of these we describe some of the background to Cauchy's work, including the first appearance, in a piece of work by d'Alembert, of what were later to be called the Cauchy–Riemann equations, the various uses made by d'Alembert, Euler and Laplace of functions of a complex variable, and the controversy between Laplace and Poisson about the validity of the method of imaginary substitutions for the evaluation of definite integrals. Chapter 2 is devoted to an analysis of Cauchy's 1814 memoir, and Chapter 3 describes various developments between 1814 and 1825. The 1825 memoir on definite integrals between imaginary limits is analysed in Chapter 4, together with a couple of other papers belonging to the same context.

Chapter 5 is concerned with Cauchy's development of his 'calculus of residues' in 1826–1829, published in a series of notes in his *Exercices de mathématiques*. Chapter 6 begins with some discussion of the Lagrange series, and goes on to show how Cauchy's work on this led him to his results, first on the convergence of the Maclaurin series of a function and his 'calculus of limits', and ultimately to those on power-series expansions for implicit functions; all this appeared in his first Turin memoir of 1831. Later in the same chapter we discuss his results in the second Turin memoir on the residue theorem for general simple closed contours and on his 'calculus of indices'.

In Chapter 7 we try to summarise the development of Cauchy's ideas on complex function theory over the period 1814–1831, not only in the context of the basic results of the theory, such as 'Cauchy's theorem' and the residue theorem, but also on some peripheral topics, such as his investigations concerning the roots of equations and his use of principal-value integrals. The chapter concludes with a brief evaluation of Cauchy's achievements over the whole period.

The final chapter is followed by a list of references which makes no claim to completeness. References in the text are given in forms such as Bidone [1812] or Cauchy [1831e]. In many cases, the date given is that of communication to an academy or other society, rather than that of publication, since there was often a considerable lapse of time between these events, but we cannot claim to have reached complete consistency in this matter.

I.3. In describing the details of mathematical arguments, we have often taken the liberty of modernising notation and terminology to some extent, with the aim of making the material easier for both the printer and the reader.

For instance, Cauchy sometimes distinguishes between two or more functions by using the same letter (usually f) from several different founts; in such cases we have either used other letters or introduced subscripts. We use the solidus notation a/b for fractions when it is convenient to do so, and we write $n!$ for the factorial function. What Cauchy calls a 'fonction entière' we shall call a polynomial. We use current notation for log, exp, tan, arctan, etc., and the notations Σ and Π for sums and products, respectively, when it is convenient to do so.

We shall speak of 'complex numbers' rather than 'imaginary expressions', and we shall consistently write i rather than $\sqrt{(-1)}$; we use $|z|$ for the absolute value of a complex number z, where Cauchy would speak of its 'modulus', and we shall use 'the imaginary part' rather than 'the coefficient of $\sqrt{(-1)}$'.

Again, we use the notation $\partial f/\partial x$ for partial derivatives rather than df/dx, and Fourier's notation

$$\int_a^b f(x)\,dx$$

for definite integrals, although Cauchy himself did not adopt it until about 1822. In this context it is worth noting that some early editors of Cauchy's collected works replaced Cauchy's original notation anachronistically by Fourier's in a number of places; in particular this was done in the reprinting of the 1814 memoir in volume 1 of the first series (1882), thereby reducing some of the 1825 footnotes to nonsense.

We shall occasionally use such notations as $\rightarrow$ (in limit operations), 'sup' and 'lim sup'.

It is hoped that the reader will not be seriously misled by these modifications, and that we shall have succeeded in faithfully conveying both the content of and the ideas behind the arguments we are describing.

I.4. Biographical information about Cauchy is available in such sources as Valson [1868], Klein [1926], pp. 70–4, 82–7, Terracini [1957], and Freudenthal [1971], and especially in Belhoste's full biography [1991]. Valson's book contains material no longer available, but treats Cauchy as a secular saint rather than a fallible human being, and lacks critical understanding of his mathematics.

Brief surveys of Cauchy's contributions to complex function theory can be found in Casorati [1868], Brill and Noether [1894], Osgood [1901], Lindelöf [1905], Kline [1972], pp. 626–70, Bottazzini [1986], pp. 126–89 and Bottazzini [1990], pp. xcvii–cxxxviii. A great deal of related information is also available in Grattan-Guinness [1990]. Writers in this list who take fully into account the significance of Cauchy's second Turin memoir include Casorati, Lindelöf and Grattan-Guinness.

Useful information about the eighteenth-century background to Cauchy's work can be found in Timchenko [1899], Stäckel [1900, 1901] and Truesdell [1954]; Timchenko also discusses some of Cauchy's major papers, including his [1814], [1815] and [1825b].

In the course of this work I have greatly benefited from discussions with Professor H. S. Tropp of Humboldt State University, California (who has read most of several successive versions) and with Mr S. P. Bursill-Hall. I have had the advantage of the superb facilities available in the Cambridge

University Library and in the library of St John's College, Cambridge. I also wish to express my thanks to the Royal Society for permitting access to some rare items in the Society's library, and to Professor I. Grattan-Guinness for the loan of a microfilm of Timchenko [1899].

1

The background to Cauchy's work on complex function theory

1.1. Cauchy's work on complex function theory began with his long memoir [1814] on definite integrals, presented to Classe I of the Institut de France in 1814, but not published until 1827. Before Cauchy's time a number of topics that were to become significant for complex function theory had been treated in the mathematical literature; in the present chapter we shall give an account of some of these, including integration round closed curves, what were to become known as the Cauchy–Riemann equations, and the use of imaginary substitutions (complex changes of variable) for the evaluation of integrals. We shall not discuss the use of complex numbers in algebra or their geometrical interpretation, nor shall we examine the treatment of the elementary functions of a complex variable. What we shall be interested in are the seeds of the ideas that Cauchy was to develop in laying the foundations of complex function theory.

1.2. One of the topics that has played an intermittent role in the history of complex function theory is the notion of an exact differential. It first appeared in a memoir[1] by Clairaut on the integral calculus; he showed that if a differential expression

$$P(x, y)\,dx + Q(x, y)\,dy$$

is exact (or, as he said, complete), i.e. if it is of the form

$$dz = df(x, y) = \frac{\partial f}{\partial x}\,dx + \frac{\partial f}{\partial y}\,dy$$

for some function $z = f(x, y)$, then

$$\frac{\partial P}{\partial y} = \frac{\partial Q}{\partial x}.$$

[1] Clairaut [1740], Première partie, Sections I–III, pp. 294–7.

A little later in his famous book on the figure of the Earth, he drew the conclusion[2] that the integral

$$\int (P\,dx + Q\,dy),$$

taken along a path joining two given points in the (x, y) plane, will be independent of the path if and only if $P\,dx + Q\,dy$ is a complete differential. As he put it, '$P\,dx + Q\,dy$ can be integrated without knowing the value of y'; in other words, without knowing the equation of the path. It would follow, although he does not say so explicitly, that the integral would vanish if it were taken round a closed curve. He goes on to say (Section XVII), in effect, that the equation

$$\frac{\partial P}{\partial y} = \frac{\partial Q}{\partial x}$$

is both necessary and sufficient for $P\,dx + Q\,dy$ to be a complete differential.

Euler also obtained[3] the condition

$$\frac{\partial P}{\partial y} = \frac{\partial Q}{\partial x}$$

for $P\,dx + Q\,dy$ to be a complete differential, but did not then pursue the problem any further.

1.3. Many years later d'Alembert [1768b, pp. 12–14] gave a counter-example, showing that Clairaut's condition is not always sufficient. Clairaut [1740, p. 297] had mentioned the differential expression

$$\frac{y\,dx - x\,dy}{x^2 + y^2}$$

as one that satisfied his condition. D'Alembert pointed out that its integral round a circle with centre at the origin does not vanish, its value being -2π. The moral he draws is that not only must $P\,dx + Q\,dy$ be a complete differential, but its integral must depend neither on the rectification nor on the quadrature of an oval curve; in other words, it should not depend on π or on any analogue of π for a general oval curve. Here he may perhaps be groping towards the idea that, if $df(x, y) = P\,dx + Q\,dy$, then f must be a single-valued function of position.

----

[2] Clairaut [1743], Section XVI, pp. 34–8.   [3] Euler [1740], p. 39.

An example closely related to this one appears in Cauchy's 1814 memoir (discussed later in Section 2.11).

1.4. What were to become known as the Cauchy–Riemann equations first appeared in print in an essay by d'Alembert [1752] on the resistance of fluids. At one point in this memoir he encounters the problem of finding functions $A(x, z)$, $B(x, z)$ such that the expressions

$$A\,dx + B\,dz, \quad zB\,dx - zA\,dz$$

are both complete (exact) differentials. He begins (p. 60) by looking at the easier problem posed by the equations

$$dq = M\,dx + N\,dz, \tag{1.1}$$

$$dp = N\,dx - M\,dz. \tag{1.2}$$

He adopts two different approaches. In the first of these (Proposition X, p. 60) he remarks that if solutions exist, then we must have

$$d(q + pi) = (M + Ni)(dx - i\,dz),$$

$$d(q - pi) = (M - Ni)(dx + i\,dz),$$

where $i = \sqrt{(-1)}$; he deduces that $M + Ni$ must be a function of $F + x - iz$ only, and $M - Ni$ of $G + x + iz$ only, where F and G are constants.

In his second approach (Corollary I, p. 61) his aim is to find expressions for p and q rather than for M and N. He says that if (1.1) and (1.2) both hold, then we must have

$$\frac{\partial p}{\partial z} = -\frac{\partial q}{\partial x} = -M, \frac{\partial p}{\partial x} = \frac{\partial q}{\partial z} = N; \tag{1.3}$$

by Clairaut's conditions,

$$q\,dx + p\,dz, \quad p\,dx - q\,dz,$$

are both complete differentials. By the same argument as in his first approach, it follows that $q + pi$ must be a function of $F + x - iz$ only and $q - pi$ of $G + x + iz$ only, where F and G are constants. After a brief discussion of conditions for p and q to be real, he ends up with expressions of the form

$$p = -i\xi(x - iz) + \zeta(x - iz) + i\xi(x + iz) + \zeta(x + iz),$$

$$q = \xi(x - iz) + i\zeta(x - iz) + \xi(x + iz) - i\zeta(x + iz),$$

where, as he puts it, the functions ξ and ζ 'contain no imaginary constants'. He illustrates his result by taking ξ and ζ to be cubic polynomials with real coefficients.

He makes no further use of these results in this memoir; instead he returns to his original problem, which he attacks by assuming that A and B are polynomials in (x, z) and obtaining relations between their coefficients.

Thus the equations (1.3) above, which are essentially the Cauchy–Riemann equations, have appeared briefly in a context that suggests complex function theory to a modern reader, although no real use has been made of them in this memoir.

1.5. Similar expressions to those described in the previous section appear in discussions of plane fluid flow by Euler [1757] Sections 70–73, d'Alembert [1761b] Sections I–V and Lagrange [1766] Sections 20–24. As Truesdell [1954] has pointed out, these papers are important historically in the development of fluid mechanics; on the other hand, they contribute little new to the ideas forming the background of complex function theory.

In the course of a discussion of a vibrating spring, d'Alembert obtains the general solution of the equation

$$\frac{\partial^2 y}{\partial x^2} = -\frac{\partial^2 y}{\partial t^2}$$

in the form

$$y = \Phi(x + ti) + \Delta(x - ti),$$

[1761a], pp. 11–14. A further point was made by Lagrange [1766], Section 20. Starting from the equations

$$\frac{\partial p}{\partial t} + \frac{\partial q}{\partial x} = 0,$$

$$\frac{\partial p}{\partial x} - \frac{\partial q}{\partial t} = 0,$$

he deduces that

$$\frac{\partial^2 p}{\partial t^2} = -\frac{\partial^2 p}{\partial x^2}. \tag{1.4}$$

He goes on to say that (1.4) has the integral

$$p = \varphi(t + x\mathrm{i}) + \psi(t - x\mathrm{i}),$$

where it follows from the equation

$$\frac{\partial q}{\partial t} = \frac{\partial p}{\partial x}$$

that

$$q = \mathrm{i}[\varphi(t + x\mathrm{i}) - \psi(t - x\mathrm{i})],$$

where φ and ψ are 'arbitrary functions'.

In both these papers d'Alembert's requirement that the functions 'contain no imaginary constants' has been dropped, with the result that the formulae are simpler in appearance than d'Alembert's original ones.

1.6. Similar arguments appear in other contexts. In a paper on orthogonal trajectories Euler [1768b] considers (Section 10) the pair of equations

$$dx = PR\,dp + QS\,dq,$$
$$dy = PS\,dp - QR\,dq,$$

whence

$$dx + i\,dy = (R + Si)(P\,dp - iQ\,dq),$$
$$dx - i\,dy = (R - Si)(P\,dp + iQ\,dq).$$

Supposing that $P\,dp \mp iQ\,dq$ has M as an integrating factor, he writes

$$T \pm Vi = \int M(P\,dp \mp iQ\,dq),$$

and then remarks that if $R \pm Si$ is a function of $T \pm Vi$, we can write down equations of the form

$$x + yi = \Gamma(T + Vi) + \frac{1}{2i}\Delta(T + Vi),$$
$$x - yi = \Gamma(T - Vi) - \frac{1}{2i}\Delta(T - Vi).$$

Again, in a paper on the conformal mapping of a sphere on a plane, Euler [1775b] is interested (Sections 42–44) in the equations

$$dx = p\,du + r\cos u \cdot dt,$$
$$dy = r\,du - p\cos u \cdot dt,$$

and treats them by the same method; Lagrange [1781] uses a similar argument (Sections 1–4) in the same context.

In these papers it has been shown that generalisations of the Cauchy–Riemann equations can be handled successfully, but little is contributed to the general theory. There is no mention here of the use of complex functions to generate conformal mappings of the plane into itself: nor does it seem that these papers had any direct influence on Cauchy's early work.

1.7. In the remainder of this chapter we shall be examining the use of complex changes of variable (or, in the language of the time, imaginary substitutions) for the evaluation of integrals; discussion of the validity of such devices was to play a part in stimulating Cauchy's early work on complex function theory.

In the early stages of the development of the integral calculus imaginary substitutions were used without any qualms. For instance, Johann Bernoulli [1702], starting from the equation

$$I = \frac{1}{b} \arctan \frac{z}{b} = \int_0^z \frac{dz}{b^2 + z^2},$$

made the substitution

$$z = \frac{ib(t-1)}{t+1},$$

thus obtaining

$$I = -\frac{1}{2bi} \int_1^t \frac{dt}{t} = -\frac{1}{2bi} \log t = \frac{i}{2b} \log \frac{bi+z}{bi-z},$$

one of the important relations between logarithms and inverse trigonometric functions.

Again, in one of the papers written by Euler [1747] to resolve the old dispute between Leibniz and Johann Bernoulli about the logarithms of negative numbers, we find the following argument. If φ is an arbitrary arc of a circle of radius 1 and $x = \sin \varphi$, then

$$d\varphi = \frac{dx}{(1-x^2)^{1/2}}.$$

The substitution $x = iz$ then gives

$$d\varphi = \frac{i\,dz}{(1+z^2)^{1/2}},$$

whence

$$\varphi = i \log [z + (1+z^2)^{1/2}] + C$$

$$= i \log \left[(1-x^2)^{1/2} + \frac{x}{i} \right] + C,$$

where C is a constant. Since $x = 0$ when $\varphi = 0$, we must have $C = 0$, whence

$$\varphi = i \log \left[(1-x^2)^{1/2} + \frac{x}{i} \right].$$

Euler then argues that this equation must hold for every value of φ such that $\sin \varphi = x$; the logarithm must therefore be a many-valued function.

1.8. In 1777 Euler communicated a series of nine papers to the St Petersburg Academy; they did not appear in print until 1793, ten years after Euler's death. We shall not discuss them in detail, but will content ourselves with an outline of the ideas appearing in them.

In the first paper [1777a] of the series, Euler starts from an equation of the form

$$\Delta(z) = \int Z(z) \, dz,$$

where z is allowed to take complex values $z = x + yi$. He splits Z and Δ into their real and imaginary parts, writing

$$Z = M + Ni, \quad \Delta = P + Qi,$$

so that

$$Z \, dz = (M + Ni)(dx + i \, dy).$$

We must therefore have

$$P = \int (M \, dx - N \, dy), \quad Q = \int (N \, dx + M \, dy), \tag{1.5}$$

so that $M \, dx - N \, dy$ and $N \, dx + M \, dy$ must both be complete differentials; by Clairaut's conditions, it follows that

$$\frac{\partial M}{\partial y} = -\frac{\partial N}{\partial x}, \quad \frac{\partial N}{\partial y} = \frac{\partial M}{\partial x}. \tag{1.6}$$

In this paper and its immediate successor [1777b] Euler gives some illustrative examples. In each of these he begins by evaluating $\int Z \, dz$ directly, and splits the result into real and imaginary parts; he then uses equation (1.5) to evaluate P and Q separately, thus verifying the result he obtained by the first method.

We note that equations (1.6) are just the Cauchy–Riemann equations; they appear here in a context a good deal closer to the basic ideas of complex function theory than in the cases discussed earlier (Sections 1.4–1.6).

In the remaining papers of the series, Euler no longer expresses his results in terms of differential forms in two real variables, as in [1777a, b]; instead he makes a complex change of variables; in [1777c, d], for instance, he uses the substitution $y = zi$. In the interesting paper [1777e] he repeats the general discussion of [1777a], adding the remark that P and Q satisfy the same pair of equations as M and N, i.e.

$$\frac{\partial P}{\partial x} = \frac{\partial Q}{\partial y}, \quad \frac{\partial P}{\partial y} = -\frac{\partial Q}{\partial x}.$$

He comments that the fact that these equations hold is not always immediately obvious when the expression $Z(z)$ is at all complicated. In this paper he takes as his illustrative example the integral

$$V(z) = \int \frac{z^{m-1} \, dz}{(a + bz^n)^{\lambda}}$$

for certain values of the exponent λ, paying special attention to the case $\lambda = m/n$, where $V(z)$ can be evaluated in terms of elementary functions. He then makes the substitution

$$z = v(\cos \theta + i \sin \theta),$$

where v is a real variable and θ a real constant. After some further manipulations, he concludes that

$$\int \frac{\alpha \sin m\omega + \beta \cos m\omega}{\gamma \sin n\omega + \delta \cos n\omega} \cdot \frac{d\omega}{(\sin n\omega)^{1-(m/n)}}$$

can be expressed in terms of elementary functions. He adds the remark that one would like to be able to obtain results of this kind without introducing imaginaries, but confesses that he can see no way of doing so.

The remaining papers of the series are variations on the same theme, with illustrative examples of various degrees of complexity, and using several different imaginary substitutions; in [1777h], for instance, we find the substitution

$$x = \frac{\cos \varphi + i \sin \varphi}{\cos^{1/n} n\varphi}$$

Incidentally, it was in this paper that he first used the notation i for $\sqrt{(-1)}$. In the final paper [1777k] Euler gives a general account of the technique he has built up in the course of the series.

In the first two papers [1777a, b] the Cauchy–Riemann equations play a genuine role; in the later ones, however, as he remarks in [1777e], what his technique provided is a method of finding many pairs of functions that satisfy them.

It should be emphasised that expressions of the form $\int Z(z) \, dz$ should not be thought of as being taken along paths in the complex plane. Euler treats integration throughout as simply the process inverse to differentiation.

1.9. We have one more Euler paper [1781] to examine; in this he uses an imaginary substitution to evaluate a definite integral. He begins with the expression

$$\Delta = \int_0^{\infty} x^{n-1} e^{-x} \, dx.$$

In the notation introduced by Legendre in 1810 we would write $\Delta = \Gamma(n)$. Using the substitution $x = ky$, where $k > 0$, he deduces that

$$\int_0^\infty y^{n-1} \, e^{-ky} \, dy = \Delta/k^n.$$

He then allows k to become complex, writing $k = p + qi$ and requiring that $p > 0$, and he puts

$$p = f \cos \theta, \qquad q = f \sin \theta,$$

where $f > 0$. Splitting the resulting equations into real and imaginary parts, he concludes that

$$\int_0^\infty x^{n-1} \, e^{-px} \cos qx \, dx = \frac{\Delta \cos n\theta}{f^n},$$

$$\int_0^\infty x^{n-1} \, e^{-px} \sin qx \, dx = \frac{\Delta \sin n\theta}{f^n}.$$

He draws attention to the special case

$$\int_0^\infty \frac{\cos x \, dx}{\sqrt{x}} = \int_0^\infty \frac{\sin x \, dx}{\sqrt{x}} = \sqrt{\left(\frac{\pi}{2}\right)}.$$

These expressions are often known as Fresnel's integrals.

As a limiting case of his results, he derives the equation

$$\int_0^\infty \frac{\sin x}{x} = \frac{\pi}{2},$$

remarking that this does seem to be correct, since numerical calculation gives nearly the same value for the integral. This hesitancy on Euler's part suggests that he may have had some qualms whether his method was completely trustworthy. In fact, he describes his whole argument as being quite singular, and says that, since it appears to be potentially useful, he has expounded it in some detail.

As we shall see later in this chapter, this paper is repeatedly referred to in discussions between Laplace and Poisson about the validity of methods involving imaginary substitutions, discussions that prepared the way for Cauchy's earliest work on definite integrals.

1.10. We now come to a remarkable argument[4] used by Laplace in his famous memoir [1782] on approximations to functions of large numbers.

He begins with the difference equation

$$(s + 1)y_s = y_{s+1}, \tag{1.7}$$

where s is not necessarily an integer. He tries a solution of the form

$$y_s = \int x^s \varphi(x) \, dx,$$

---

[4] Laplace [1782], pp. 53–60; *Oeuvres*, **10**, 258–64.

where the limits of integration are to be chosen later. Substituting this expression in (1.7), and performing an integration by parts, he obtains

$$\int x^s \left[(1-x)\varphi - \frac{\mathrm{d}}{\mathrm{d}x}(x\varphi) \right] \mathrm{d}x + [\varphi(x)x^{s+1}] = 0.$$

This can be satisfied if

$$(1-x)\varphi - \frac{\mathrm{d}}{\mathrm{d}x}(x\varphi) = 0$$

and $\varphi(x)x^{s+1}$ vanishes at the limits of integration. The differential equation gives

$$\varphi(x) = A\,\mathrm{e}^{-x},$$

where A is a constant, and the other condition can then be satisfied by taking the range of integration as $(0, \infty)$, provided that $s + 1 > 0$.

He has thus obtained

$$y_s = A \int_0^\infty x^s\,\mathrm{e}^{-x}\,\mathrm{d}x$$

as a solution of the difference equation (1.7).

Laplace now supposes it is known that $y_s = Y$ when s takes the particular value $s = \mu$. By methods that he has developed earlier in the memoir, he obtains an asymptotic expansion of y_s for large values of s; this is of the form

$$y_s \sim \frac{Ys^{s+1/2}\,\mathrm{e}^{-s}\sqrt{(2\pi)}}{\displaystyle\int_0^\infty x^\mu\,\mathrm{e}^{-x}\,\mathrm{d}x} \left(1 + \frac{1}{12s} + \cdots \right). \tag{1.8}$$

On the other hand, if $s - \mu$ is a positive integer, equation (1.7) has the obvious solution

$$y_s = Y(\mu + 1)(\mu + 2)\ldots(s - 1)s. \tag{1.9}$$

Eliminating Y from (1.8) and (1.9), he obtains

$$(\mu + 1)(\mu + 2)\ldots(s - 1)s \sim \frac{s^{s+1/2}\,\mathrm{e}^{-s}\sqrt{(2\pi)}}{\displaystyle\int_0^\infty x^\mu\,\mathrm{e}^{-x}\,\mathrm{d}x} \left(1 + \frac{1}{12s} + \cdots \right). \tag{1.10}$$

Laplace now alleges that (1.8) will still hold if s and μ are negative, except that one may have to use different limits of integration in the denominator. He replaces s and μ in (1.8) by $-s$ and $-\mu$, obtaining

$$y_{-s} \sim \frac{Y\mathrm{i}\,\mathrm{e}^s\sqrt{(2\pi)}}{(-1)^s s^{s-1/2}\displaystyle\int x^{-\mu}\,\mathrm{e}^{-x}\,\mathrm{d}x} \left(1 - \frac{1}{12s} + \cdots \right), \tag{1.11}$$

where Y is now the value of y_s when $s = -\mu$. To ensure that the integrand in the denominator vanishes at the limits of integration, he makes the substitution

$$x = -\mu + \varpi i$$

and takes the integral over the range $-\infty < \varpi < \infty$. After a little reduction, the integral becomes

$$\frac{i\,e^{\mu}}{(-1)^{\mu}} \int_0^{\infty} \left\{ \frac{e^{-\varpi i}}{(\mu - \varpi i)^{\mu}} + \frac{e^{\varpi i}}{(\mu + \varpi i)^{\mu}} \right\} d\varpi = \frac{i\,e^{\mu}}{(-1)^{\mu}} \int_0^{\infty} Q(\varpi)\,d\omega,$$

say. Substituting this expression in (1.11), he obtains

$$y_{-s} \sim \frac{Y\,e^{s-\mu}\sqrt{(2\pi)}}{(-1)^{s-\mu}s^{s-1/2} \int_0^{\infty} Q(\varpi)\,d\varpi} \left(1 - \frac{1}{12s} + \cdots \right). \qquad (1.12)$$

On the other hand, if $s - \mu$ is a positive integer, the difference equation (1.7) gives directly

$$y_{-s} = \frac{(-1)^{s-\mu}Y}{\mu(\mu + 1)\ldots(s - 2)(s - 1)}. \qquad (1.13)$$

Eliminating Y from (1.12) and (1.13) then gives

$$(\mu + 1)(\mu + 2)\ldots(s - 1)s \sim \frac{s^{s+1/2}\,e^{\mu-s}}{\mu\sqrt{(2\pi)}} \int_0^{\infty} Q\,d\varpi \left(1 + \frac{1}{12s} + \cdots \right). \qquad (1.14)$$

Equating the right-hand sides of (1.10) and (1.14), Laplace concludes that

$$\int_0^{\infty} Q(\varpi)\,d\varpi = \frac{2\pi\mu\,e^{-\mu}}{\displaystyle\int_0^{\infty} x^{\mu}\,e^{-x}\,dx}. \qquad (1.15)$$

This result is in fact correct, in spite of the unusual character of the argument.

He examines some special cases; thus, if $\mu = 1$, (1.15) leads to the equation

$$\int_0^{\infty} \frac{\cos\varpi + \varpi\sin\varpi}{1 + \varpi^2}\,d\varpi = \frac{\pi}{e}.$$

The passage we have discussed here is the only one in this memoir in which an imaginary substitution occurs. It was on the basis of this piece of work that Laplace was later to claim priority for the use of imaginary substitutions as a means to the evaluation of definite integrals, making the point that it was published in 1785, whereas Euler [1781] did not appear in print until 1794.

1.11. After the appearance of Laplace's paper [1782] the method of imaginary substitutions seems to have been neglected for some time. Its revival by Laplace himself in 1809 sparked off a lively exchange of views between him and Poisson; this will occupy us for most of the remainder of the present chapter.

The story begins with Laplace's paper [1809], which was in certain respects a sequel to his paper [1782]. He begins the third section of this paper, entitled 'Sur le passage réciproque des résultats réels aux résultats imaginaires', by remarking that when results are expressed in terms of variables (he uses the phrase 'quantités indéterminées') the generality of the notation covers both real and imaginary cases, adducing as an example the exponential representation of the trigonometric functions. He then adds that he showed in his paper [1782] that one can also use this passage from real to imaginary when the results are expressed in terms of constant parameters (he calls them 'quantités déterminées'); this had enabled him to deduce the values of some definite integrals that would be difficult to obtain in any other way. He now proposes to give some new applications of this remarkable artifice.

In the integral

$$I = \int_0^\infty x^{-\alpha}\, e^{ix}\, dx,$$

where $0 < \alpha < 1$, he makes the substitution $x = i t^{1/(1-\alpha)}$, obtaining

$$I = \frac{i^{1-\alpha}}{1-\alpha} \int_0^\infty e^{-t^{1/(1-\alpha)}}\, dt = \frac{k i^{1-\alpha}}{1-\alpha},$$

where k is a real constant depending on α. After a long discussion of the correct value of $i^{1-\alpha}$ to be chosen, he concludes that

$$\int_0^\infty x^{-\alpha} \cos x\, dx = \frac{k}{1-\alpha} \cos\tfrac{1}{2}\pi(1-\alpha),$$

$$\int_0^\infty x^{-\alpha} \sin x\, dx = \frac{k}{1-\alpha} \sin\tfrac{1}{2}\pi(1-\alpha).$$

He is clearly relying on the principle of the generality of analysis[5] to justify his argument.

In his paper [1810] Laplace again uses some imaginary substitutions. His comments on the use of this technique are somewhat fuller than in Laplace [1809], and deserve quotation. He says:

Ces approximations se déduisent encore très simplement du passage réciproque des résultats imaginaires aux resultats réels, dont j'ai donné divers exemples dans les Mémoires cités[6] .... Il est analogue à celui des nombres entiers positifs aux

---

[5] See, for example, Smithies [1986].   [6] [1782] and [1809].

nombres négatifs et aux nombres fractionnaires, passage dont les géomètres ont su tirer, par induction, beaucoup d'importants théorèmes; employé comme lui avec réserve, il devient un moyen fécond de découvertes, et il montre de plus en plus la généralité de l'analyse.'

One gets the impression that Laplace is here a little on the defensive; his introduction of an analogy with the passage from results about the positive integers to results about negative and fractional numbers suggests that he feels the need to make a case for his procedure. It is possible that Poisson, whose first published comments we discuss in the next section, had already been expressing some scepticism about its validity.

1.12. In 1811 Poisson published the note [1811a] on definite integrals; in this he comments on Laplace's paper [1810], which we described in the previous section. He says:

M. Laplace a donné des intégrales définies de différentes formules qui contiennent des sinus ou des cosinus. Il les a déduites des intégrales des exponentielles, par une sorte d'induction fondée sur le passage des quantités réelles aux imaginaires.

He then proceeds to obtain and generalise some of Laplace's results by an alternative method, based on changing the order of integration in double integrals, thus avoiding any recourse to imaginary substitutions, and remarks that Laplace had himself used this device.

Laplace had actually used the device to evaluate the integral

$$\int_0^\infty e^{-x^2} \, dx$$

in the following way.[7] We have

$$I = \int_0^\infty \int_0^\infty e^{-s(1+x^2)} \, ds \, dx = \int_0^\infty \frac{dx}{1+x^2} = \frac{\pi}{2}.$$

We also have

$$I = \int_0^\infty e^{-s} \, ds \int_0^\infty e^{-sx^2} \, dx$$

$$= \int_0^\infty s^{-1/2} e^{-s} \, ds \int_0^\infty e^{-x^2} \, dx$$

$$= 2 \int_0^\infty e^{-s^2} \, ds \int_0^\infty e^{-x^2} \, dx$$

$$= 2 \left(\int_0^\infty e^{-x^2} \, dx \right)^2.$$

[7] Laplace [1782], pp. 13–14; *Oeuvres*, **10**, 222. See also Laplace [1780], pp. 292–3; *Oeuvres*, **9**, 447.

It follows that

$$\int_0^\infty e^{-x^2}\,dx = \tfrac{1}{2}\sqrt{\pi}.$$

Poisson also evaluates some definite integrals by obtaining and solving a differential equation with respect to a parameter appearing in the integrand.

Poisson's description in this paper of Laplace's method as a kind of induction appears as a persistent theme throughout the subsequent discussion.

1.13. In the next number of the same periodical, Laplace published a short note [1811a], in which he uses the double-integral method to obtain the formulae

$$\int_0^\infty \frac{\cos ax}{1+x^2}\,dx = \int_0^\infty \frac{x\sin ax}{1+x^2}\,dx = \tfrac{1}{2}\pi e^{-a} \quad (a>0),$$

thus attaining much more directly some of the results of his 1782 memoir (Section 1.10 above).

This note, which does not appear in Laplace's collected works, is an extract from the longer memoir [1811b], published in the same year. In his introduction to the latter, he refers to the method of imaginary substitutions in language similar to that used in Laplace [1810], but he adds the remark

... ces moyens, comme celui de l'induction, quoique employés avec beaucoup de précaution et de réserve, laissent toujours à désirer des démonstrations directes de leurs résultats.

He mentions the results that Poisson had obtained in his paper [1811a] and says that in the present paper he will give a direct derivation of these and some more general results.

He does in fact deal with the integral

$$\int_0^\infty x^{-\omega} e^{-ax-irx}\,dx,$$

where $\omega < 1$ and $a > 0$, which was evaluated by Euler [1781] by using imaginary substitutions; Laplace's method here is to expand e^{irx} as a power series in x and integrate term by term. Later in the paper, however, he uses imaginary substitutions to evaluate a number of integrals without suggesting any alternative arguments. Thus, in Section III, he repeats what he had done in Laplace [1811a], but then, in order to evaluate

$$\int_0^\infty \frac{\cos rx}{1+\beta^2 x^2}\,dx,$$

where β may be imaginary, he makes the substitution $\beta x = x'$. .

We note that, although Laplace has accepted the need for caution in the use of imaginary substitutions, and has admitted the desirability of verifying the results by more direct methods, he is not prepared to abandon the technique altogether.

Poisson published an abstract [1811b] of this memoir later in the same year. In commenting on the argument used by Laplace in his Section III, he insists on the importance of confirming results obtained in this way by a more direct method, and promises to deal directly with these integrals in a later paper. This promise was fulfilled in his paper [1813] (see Section 1.15 below).

1.14. In Laplace's famous book *Théorie analytique des probabilités* [1812, 1814b, 1820] there are several passages concerned with imaginary substitutions.

For instance, we find the following statement:[8]

Les limites des intégrales définies que cette méthode réduit en séries convergentes sont ... données par les racines d'une équation que l'on peut nommer *équation des limites*. Mais une remarque trés importante dans cette analyse ... est que les séries auxquelles on parvient ont également lieu dans le cas même où, par des changements de signe dans les coefficients de l'équation des limites, ces racines deviennent imaginaires.

He then repeats the passage from Laplace [1811b] that we quoted in Section 1.13, including the remark about the desirability of direct proofs. He continues:

Leur rapprochement [i.e. ces moyens] des méthodes directes servent à les confirmer et à faire voir la grande généralité de l'analyse ...; j'ai insisté sur ces passages qu'Euler considérait en même temps que moi, et dont il a fait plusieurs applications curieuses, mais qui n'ont paru que depuis la publication des Mémoires cités.

He is clearly referring here to Euler [1781] and Laplace [1782].

One example of a use of imaginary substitutions[9] in this book is worth describing. Having just proved that

$$\int_0^\infty e^{-t^2}\, dt = \tfrac{1}{2}\sqrt{\pi},$$

Laplace goes on to say that, in virtue of the generality of analysis, the formula can be extended to cases in which t is imaginary. He then argues as follows; we have

$$\int_0^\infty e^{-a^2 x^2} \cos rx\, dx = \tfrac{1}{2}\int_0^\infty e^{-a^2 x^2 + rxi}\, dx + \tfrac{1}{2}\int_0^\infty e^{-a^2 x^2 - rxi}\, dx.$$

---

[8] *Oeuvres*, **7**, 87–8.   [9] *Oeuvres*, **7**, 96–7.

The first integral on the right-hand side can be rewritten as

$$e^{-r^2/4a^2} \int_0^\infty e^{-(ax-ri/2a)^2} \, dx,$$

and the substitution $t = ax - (ri/2a)$ reduces it to

$$a^{-1} e^{-r^2/4a^2} \int_{-ri/2a}^\infty e^{-t^2} \, dt.$$

Similarly, the second integral becomes

$$a^{-1} e^{-r^2/4a^2} \int_{ri/2a}^\infty e^{-t^2} \, dt = a^{-1} e^{-r^2/4a^2} \int_{-\infty}^{-ri/2a} e^{-t^2} \, dt.$$

The two integrals together give

$$a^{-1} e^{-r^2/4a^2} \int_{-\infty}^\infty e^{-t^2} \, dt = a^{-1} \sqrt{\pi}\, e^{-r^2/4a^2},$$

and he concludes that

$$\int_0^\infty e^{-a^2 x^2} \cos rx \, dx = (\sqrt{\pi}/2a) e^{-r^2/4a^2}.$$

From a modern point of view, what he has done here is equivalent to moving the contour of integration from the real axis to a line parallel to it and passing through the point $-ri/2a$.

Laplace gives an alternative proof of the result by obtaining a differential equation for the integral as a function of r and solving it directly. We recall that Poisson had already used this technique (Section 1.12 above).

Various other imaginary substitutions are scattered through the book; in a few cases Laplace does provide an alternative proof. Similar devices also occur in passages added in the second and third editions.

We note that Laplace seems to be recovering his confidence in the method of imaginary substitutions, many of his results having been confirmed by other methods, although he still admits that such confirmation is desirable; we note also that he invokes Euler's authority in his support, while maintaining his claim to priority of publication.

1.15. In his paper [1813] Poisson carried out the promise he had made in his earlier paper [1811b] (Section 1.13 above); this paper was the first of a series on definite integrals. He uses his differential equation method to obtain the results of Euler [1781] (Section 1.9 above), and remarks (p. 219):

Ces formules sont dues à Euler, qui les a trouvées par une sorte d'induction fondée sur le passage des quantités réelles aux imaginaires; induction qu'on peut bien employer comme un moyen de découverte, mais dont les résultats ont besoin d'être confirmés par des méthodes directes et rigoureuses.

He goes on to evaluate a number of the integrals obtained by Laplace, and some new ones as well.

We note that Poisson is here insisting even more strongly than before on the necessity of direct confirmation of results obtained by the use of imaginary substitutions.

1.16. Laplace returned to the topic of imaginary substitutions in 1814, by which time he seems to have shifted his position just a little. In his *Essai philosophique sur les probabilités* [1814a] we find the following passage (*Oeuvres*, **7**, xl):

On peut donc considérer ces passages comme un moyen de découverte, pareil à l'induction et à l'analyse, employées longtemps par les géomètres, d'abord avec une extrême réserve, ensuite avec une entière confiance, un grand nombre d'exemples en ayant justifié l'emploi. Cependant il est toujours nécessaire de confirmer par des démonstrations directes les résultats obtenus par ces divers moyens.

Again, in a section added in the second edition of his *Théorie analytique* [1814b], we find (*Oeuvres*, **7**, 480):

Tous ces moyens d'invention, qui tiennent à la généralité de l'Analyse, exigent dans leur usage une grande circonspection, et il est toujours bon d'en démontrer directement.

He appears to be more cautious here than he had been in 1812 (Section 1.14 above). The first passage quoted here echos some of Poisson's remarks: it is also possible that he had been discussing the matter with Cauchy, whose memoir on definite integrals, containing a completely new approach to the problem, was submitted to the Première Classe of the Institut de France in August 1814 (see Chapter 2 below).

1.17. Methods involving imaginary substitutions also appeared elsewhere. For instance, on p. 100 of Part 4 of Legendre's *Exercices du calcul intégral* [Legendre 1814a], he starts from the formula

$$\int_0^\infty \frac{z^{a-1}\,dz}{(1+z)^r} = \frac{\Gamma(a)\Gamma(r-a)}{\Gamma(r)} \qquad (0 < a < r);$$

replacing z by kz, he obtains

$$\int_0^\infty \frac{z^{a-1}\,dz}{(1+kz)^r} = k^{-a}\frac{\Gamma(a)\Gamma(r-a)}{\Gamma(r)};$$

he then takes $k = c(\cos\theta + i\sin\theta)$. One of the results he arrives at in this way is

$$\int_0^\infty \frac{z^a\,dz}{1 + 2cz\cos\theta + c^2 z^2} = \frac{\pi c^{-a}}{\sin a\pi} \cdot \frac{\sin a\theta}{\sin\theta} \qquad (0 < a < 1).$$

1.18. Let us now summarise briefly what had been achieved up to 1814 in directions foreshadowing complex function theory.

The notion of a complete (exact) differential had been introduced by Clairaut, and d'Alembert had drawn attention to the possibility that the integral of a complete differential round a closed curve might not vanish.

The Cauchy–Riemann equations and generalisations of them had appeared in various contexts involving functions of a complex variable, including fluid mechanics, orthogonal trajectories, and conformal mapping, and their relation to the two-dimensional Laplace equation had been recognised.

Imaginary substitutions had been used in the integral calculus from the time of Johann Bernoulli onwards, at first quite freely and without hesitation. Their use for the evaluation of definite integrals had been introduced by Euler and Laplace. Euler had remarked on the unusual character of this type of argument. Laplace at first used it without comment; after Poisson had expressed some doubts about its validity, Laplace sought justification for the method in the principle of the generality of analysis, but he eventually came to admit that results obtained in this way ought to be confirmed by more direct proofs.

This was the situation when Cauchy was preparing his memoir [1814] on definite integrals, which, although submitted to the Première Classe of the Institut de France in 1814, was not actually published until 1827. Belhoste [1991, pp. 107–8] conjectures very plausibly that an examination of the method of imaginary substitutions was suggested to Cauchy by Laplace, in order to resolve the controversy between himself and Poisson.

We shall analyse Cauchy's 1814 memoirs in detail in the next chapter.

2

Cauchy's 1814 memoir on definite integrals

2.1. Cauchy's contributions to the development of complex function theory began with his long memoir [1814] on definite integrals; it was presented to Classe I of the Institut de France on 22 August 1814, the day after Cauchy's 25th birthday. The memoir was examined by Legendre and Lacroix, and its publication was recommended in their report [Legendre 1814b], written by Legendre, and dated 7 November 1814. The original intention was that the memoir should appear in the *Mémoires présentés à l'Institut des Sciences, Lettres et Arts, par divers savans*, but no further issues of this periodical were published because of the reorganisation of the Institut after the the restoration of the Bourbon dynasty. The first volume of the replacement periodical, the *Mémoires présentés par divers savans à l'Académie Royale des Sciences de l'Institut de France*, did not appear until 1827, and did contain both this memoir and Cauchy's prize essay [1815] on the theory of waves.

Because of the long delay in the publication of the 1814 memoir. Cauchy summarised some of its results in various publications between 1814 and 1825 (see Chapter 3 below). By 1823 he was clearly wondering whether it would ever appear in the Academy's publications; in [1823a] he announced that it would probably appear in the next number of the *Journal de l'Ecole Polytechnique*, but this too was delayed, not being published till 1831, so this scheme did not work. The manuscript was ultimately sent for printing on 14 September 1825; in its final form Cauchy inserted numerous and extensive footnotes, apparently leaving the main text unaltered. In the present chapter we shall ignore these footnotes, postponing their discussion to Chapter 3.

2.2. In his introduction to the 1814 memoir Cauchy remarks that the solution of many problems reduces to the evaluation of definite integrals, many curious and useful results having been obtained by Euler, Laplace and Legendre. Some of the evaluations of integrals by Euler and Laplace were, he says, first discovered by a kind of induction based on a passage from real to imaginary. He refers to the statement made by Laplace [1812] and described in Section 1.14 above, quoting the passage where Laplace defends such methods as a means of discovering results but admits the desirability of more direct proofs. Cauchy goes on to say that Laplace took care to confirm by other methods the results he obtained in this way; in fact, as we have seen in Chapter 1, although Laplace gave alternative proofs for some of his results, he did not do so for all of them. Cauchy then mentions that Poisson ([1811a] and [1813]) had performed a similar service for Euler, using methods involving double integrals or the solution of differential equations.

Cauchy announces his aims in the 1814 memoir in the following words:

Après avoir réfléchi sur cet objet, et rapproché les uns des autres les divers résultats ci-dessus mentionnés, j'ai conçu l'espoir d'établir le passage du réel à l'imaginaire sur une analyse directe et rigoureuse; et mes recherches m'ont conduit à la méthode qui fait l'objet de ce Mémoire.

The remainder of the introduction is devoted to a summary of the contents of the memoir; this is couched in almost purely verbal terms, avoiding the introduction of any formal equations.

The memoir is divided into two Parts; in modern language, as we shall see, these are devoted, respectively, to results equivalent to certain special cases of 'Cauchy's theorem' and to results equivalent to special cases of the residue theorem.

2.3. Part I of the 1814 memoir is entitled 'Des équations qui autorisent le passage du réel à l'imaginaire'. This title refers in fact to the phraseology used by Laplace and Poisson (see Chapter 1 above) and not directly to what Cauchy does in the present paper, where he systematically splits almost every complex equation into its real and imaginary parts; in the introduction to his *Analyse algébrique* [1821a, p. iv] he was to insist that every complex equation should be interpreted as representing two equations between real quantities. His adoption of this expository style in the present memoir may also have been intended to demonstrate how rigorous his arguments were. In fact, he was to abandon this practice within a few years (see Section 3.3 below).

He lays the foundations of his general method in Section I of Part I. His starting-point is an 'arbitrary' function $f(y)$ of a variable y, which is in turn assumed to be a function of two real variables x and z. If $\int f(y)\,dy$ is an indefinite integral of $f(y)$, then we must have

$$\frac{\partial}{\partial x}\int f(y)\,dy = f(y)\frac{\partial y}{\partial x},$$

$$\frac{\partial}{\partial z}\int f(y)\,dy = f(y)\frac{\partial y}{\partial z}.$$

He differentiates these equations with respect to z and x, respectively; taking for granted the equality of the mixed partial derivatives, he obtains

$$\frac{\partial}{\partial z}\left[f(y)\frac{\partial y}{\partial x}\right] = \frac{\partial}{\partial x}\left[f(y)\frac{\partial y}{\partial z}\right]. \tag{2.1}$$

He also verifies this equation directly.

We now come to the crucial point of Cauchy's argument; he presumes that equation (2.1) will still hold if y, as a function of x and z, is partly real and partly imaginary. He writes y in the form

$$y = M + Ni,$$

where M and N are real-valued functions of x and z, and then puts[1]

$$f(y) = f(M + Ni) = P + Qi,$$

where P and Q are real. We then have

$$f(y)\frac{\partial y}{\partial x} = \left(P\frac{\partial M}{\partial x} - Q\frac{\partial N}{\partial x}\right) + \left(P\frac{\partial N}{\partial x} + Q\frac{\partial M}{\partial x}\right)i$$

$$= S + Ti,$$

say, where S and T are real. Similarly,

$$f(y)\frac{\partial y}{\partial z} = \left(P\frac{\partial M}{\partial z} - Q\frac{\partial N}{\partial z}\right) + \left(P\frac{\partial N}{\partial z} + Q\frac{\partial M}{\partial z}\right)i$$

$$= U + Vi,$$

where U and V are real. Equation (2.1) then becomes

$$\frac{\partial S}{\partial z} + \frac{\partial T}{\partial z}i = \frac{\partial U}{\partial x} + \frac{\partial V}{\partial x}i.$$

He seems to feel that he requires a special argument to justify equating real and imaginary parts in this equation; what he does is to assume tacitly that we also have

$$f(M - Ni) = P - Qi,$$

whence a repetition of the above argument gives

$$\frac{\partial S}{\partial z} - \frac{\partial T}{\partial z}i = \frac{\partial U}{\partial x} - \frac{\partial V}{\partial x}i.$$

[1] For simplicity of notation, we write P, Q for Cauchy's P', P''.

Combining this with his previous result, he deduces the final pair of equations

$$\frac{\partial S}{\partial z} = \frac{\partial U}{\partial x}, \quad \frac{\partial T}{\partial z} = \frac{\partial V}{\partial x}. \tag{2.2}$$

The same device was used by Euler, for instance in [1777e]. He then remarks that these equations can also be verified directly.

At this point he asserts that the equations (2.2) contain the whole theory of the passage from the real to the imaginary, and that all we have to do now is to show how they can be used.

We pause for a moment to note that the 'Cauchy–Riemann' equations are a special case of (2.2); for, if we take $y = x + zi$, so that $M = x$, $N = z$, then we have

$$S = P,\, T = Q,\, U = -Q,\, V = P,$$

and equations (2.2) become

$$\frac{\partial P}{\partial z} = -\frac{\partial Q}{\partial x}, \quad \frac{\partial Q}{\partial z} = \frac{\partial P}{\partial x}.$$

Cauchy's next move is to multiply equations (2.2) by $dx\,dz$ and then to integrate with respect to x and z between fixed (real) limits. The resulting equations will be easier to understand if we modify Cauchy's notation somewhat, writing (x_0, X) and (z_0, Z) for the limits of integration (as Cauchy often did later on); we then obtain

$$\left. \begin{array}{l} \displaystyle\int_{x_0}^{X} S(x,\, Z)\,dx - \int_{x_0}^{X} S(x,\, z_0)\,dx = \int_{z_0}^{Z} U(X,\, z)\,dz - \int_{z_0}^{Z} U(x_0,\, z)\,dz \\[3mm] \displaystyle\int_{x_0}^{X} T(x,\, Z)\,dx - \int_{x_0}^{X} T(x,\, z_0)\,dx = \int_{z_0}^{Z} V(X,\, z)\,dz - \int_{z_0}^{Z} V(x_0,\, z)\,dz. \end{array} \right\} \tag{2.3}$$

Everything in Part I of the memoir is ultimately derived from these fundamental equations.

At this point some comments on Cauchy's argument are in order. In the first place, he is clearly assuming that the functions he is considering can be differentiated as often as is necessary, even with respect to a complex variable. Behind this lies the usual eighteenth-century concept of a function as being given by an analytical expression; such functions usually had derivatives except at isolated singular points. Even after introducing his definition of continuity in the *Analyse algébrique* in 1821, he tacitly assumes that continuous functions have the same property, even when they are functions of a complex variable; his main modification of the older ideas, as will appear later in the present chapter, is that he accepts the possibility that a function may be represented by different analytic expressions in different intervals. It was not until 1839 that he occasionally begins to require a function to have a continuous derivative (Section 6.9 below), and he does not appear to appreciate the intrinsic significance of the possession of a derivative by a function of a complex variable until about 1851.

One minor point: his assumption that if $f(M + Ni) = P + Qi$ then $f(M - Ni) = P - Qi$ implies that $f(y)$ must be real for real values of y. This hypothesis appears from time to time in later papers, but it eventually disappears.

2.4. Let us now examine what equations (2.3) of Section 2.3 signify in modern terms. In geometrical language, which Cauchy scrupulously avoids using in this memoir, he has obtained these equations by integrating equations (2.2) over the rectangle $x_0 \leqslant x \leqslant X$, $z_0 \leqslant z \leqslant Z$. We can now combine the two equations to obtain

$$\int_{x_0}^{X} [S(x, Z) + iT(x, Z)]\, dx - \int_{x_0}^{X} [S(x, z_0) + iT(x, z_0)]\, dx$$

$$= \int_{z_0}^{Z} [U(X, z) + iV(X, z)]\, dz - \int_{z_0}^{Z} [U(x_0, z) + iV(x_0, z)]\, dz. \tag{2.4}$$

To see more clearly what is going on here, we first look again at the special case $y = M + Ni = x + zi$, so that $S = P$, $T = Q$, $U = -Q$, $V = P$. Equation (2.4) then becomes

$$\int_{x_0}^{X} [P(x, Z) + iQ(x, Z)]\, dx - \int_{x_0}^{X} [P(x, z_0) + iQ(x, z_0)]\, dx$$

$$= i\int_{z_0}^{Z} [P(X, z) + iQ(X, z)]\, dz - i\int_{z_0}^{Z} [P(x_0, z) + iQ(x_0, z)]\, dz,$$

which we can write more compactly as

$$\int_{x_0}^{X} f(x + iZ)\, dx - \int_{x_0}^{X} f(x + iz_0)\, dx = i\int_{z_0}^{Z} f(X + iz)\, dz - i\int_{z_0}^{Z} f(x_0 + iz)\, dz;$$

in other words, the difference between the integrals of $f(x + iz)$ along the upper and lower sides of the rectangle is equal to i times the difference between the integrals of $f(x + iz)$ along the right-hand and left-hand sides of the rectangle. We immediately recognise that this is just 'Cauchy's theorem' for the special case of a rectangle.

In the general case, we have

$$y = M + Ni,$$

M and N being functions of (x, z). We can then think of the equations

$$u = M(x, z), \quad v = N(x, z)$$

as defining a mapping of the (x, z)-plane into the (u, v)-plane. The lines x = constant, z = constant, in the (x, z)-plane, will map into two families of curves in the (u, v)-plane, so that the sides of the rectangle $x_0 \leqslant x \leqslant X$, $z_0 \leqslant z \leqslant Z$ will in general map into a curvilinear quadrilateral ABDC in the (u, v)-plane, with the arcs AB, CD, AC, BD lying on the images of the lines $z = z_0$, $z = Z$, $x = x_0$, $x = X$, respectively (Fig. 2.1). Recalling that

$$S + iT = f(y)\frac{\partial y}{\partial x}, \quad U + iV = f(y)\frac{\partial y}{\partial z},$$

we see that equation (2.4) becomes

$$\int_{CD} f(y)\,dy - \int_{AB} f(y)\,dy = \int_{BD} f(y)\,dy - \int_{AC} f(y)\,dy.$$

In other words, we have obtained 'Cauchy's theorem' for the curvilinear[2] quadrilateral ABDC in the (u, v)-plane.

What emerges from this discussion is that at this stage Cauchy is already in possession of a powerful tool that he can use to obtain many of the results for which today we should employ the full extent of Cauchy's theorem as we know it. The form in which Cauchy has expressed his results here does, however, tend to obscure the essential features of the situation.

To what extent Cauchy thought of these results in geometrical terms we do not know; there is no clear evidence either way, partly because the fashion of the time was to express everything in purely analytical terms, as is witnessed by Lagrange's boast that there were no diagrams in his *Mécanique analytique*. Cauchy's first published use of geometrical language in this context was in his 1825 memoir (Section 4.8 below). There is some evidence that in 1814 he did not yet have a clear understanding of what was going on (Section 2.12 below).

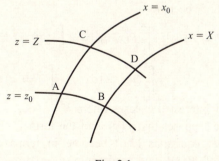

Fig. 2.1

---

[2] This interpretation is well described by Ettlinger [1923].

2.5. In the next section (Section II) of Part I Cauchy gives his first applications of the basic equations (2.3) of Section 2.3 above. He takes the simplest case $y = x + zi$, so that

$$S = V = P, \quad T = -U = Q,$$

and also supposes that $x_0 = z_0 = 0$. The equations then reduce to

$$\left.\begin{aligned}
\int_0^X P(x,\, Z)\, dx - \int_0^X P(x,\, 0)\, dx &= -\int_0^Z Q(X,\, z)\, dz + \int_0^Z Q(0,\, z)\, dz, \\
\int_0^X Q(x,\, Z)\, dx &= \int_0^Z P(X,\, z)\, dz - \int_0^Z P(0,\, z)\, dz.
\end{aligned}\right\} \quad (2.5)$$

The lack of symmetry in these equations arises from his tacit assumption that $f(y)$ is real for real y, which implies that $Q(x,\, 0) = 0$.

As a first example, he takes $f(y) = e^{-y^2}$ and $X = +\infty$, obtaining

$$\int_0^\infty e^{-x^2} \cos 2xz\, dz = e^{-z^2} \int_0^\infty e^{-x^2}\, dx = \tfrac{1}{2}\sqrt{\pi} \cdot e^{-z^2},$$

$$\int_0^\infty e^{-x^2} \sin 2xz\, dz = e^{-z^2} \int_0^z e^{-z^2}\, dz.$$

These were already known [Legendre 1811, pp. 362–3].

He then looks at a number of special situations. He remarks, for instance, that if $Q(0,\, z) = Q(X,\, z) = 0$, the first equation of (2.5) takes the simple form

$$\int_0^X P(x,\, Z)\, dx = \int_0^X P(x,\, 0)\, dx;$$

in many cases, he says, the value of X that makes $Q(X,\, z) = 0$ is $X = +\infty$, as in the example above.

Another special situation he considers is given by

$$P(X,\, z) = Q(X,\, z) = 0,$$

so that equations (2.5) become

$$\left.\begin{aligned}
\int_0^X P(x,\, Z)\, dx - \int_0^X P(x,\, 0)\, dx &= \int_0^Z Q(0,\, z)\, dz, \\
\int_0^X Q(x,\, Z)\, dx &= -\int_0^Z P(0,\, z)\, dz.
\end{aligned}\right\} \quad (2.6)$$

Again he is mainly interested in cases where $X = +\infty$.

In particular, he supposes that $f(y)$ is of the form $y^{n-1} F(y)$, and works out in detail what equations (2.6) become in terms of the real and imaginary parts of $F(y)$; this leads him to a group of four rather complicated formulae. As an illustration, he takes $F(y) = e^{-y^2}$, and evaluates the integrals

$$\int_0^\infty x^{2k}\, e^{-x^2} \cos ax\, dx, \quad \int_0^\infty x^{2k-1} e^{-x^2} \sin ax\, dx$$

for positive integral values of k. He claims that this method gives more explicit expressions for these integrals than those obtained by Legendre [1811, p. 363] by repeated differentiation of the integral

$$\int_0^\infty e^{-x^2} \cos ax\, dx = \tfrac{1}{2}\sqrt{\pi}\, e^{-a^2/4}$$

with respect to a.

Throughout this section, he tacitly assumes that if $P(X, z)$ or $Q(X, z)$ vanishes when $X = +\infty$, then $\int_0^z P(X, z)\, dz$ or $\int_0^z Q(X, z)\, dz$ will also vanish when $X = +\infty$. This has the effect that the conditions he imposes on his functions for large values of the variable are insufficient to guarantee the convergence of the infinite integrals he introduces. Cauchy continued to make such assumptions tacitly for many years; no trouble arises here, nor indeed anywhere in the 1814 memoir, but at a later stage, as we shall see, he was led to some erroneous results (Sections 3.9 and 4.14 below), and he ultimately found it necessary to impose more stringent conditions (Sections 5.5 and 5.10 below).

2.6. In the next three sections (Sections III–V) Cauchy considers the effects of some other choices for the functions M and N.

In Section III he takes

$$y = M + Ni = ax + xzi.$$

In the representation described in Section 2.4 above, the lines $x = $ const. map into lines $u = $ const. parallel to the v-axis, and the lines $z = $ const. map into lines $v/u = $ const., i.e. lines through the origin in the (u, v)-plane.

We shall not detail the formulae that Cauchy obtains in this case; we need only remark that what he is doing, in the case $X = +\infty$, is equivalent, in modern terms, to transforming

$$\int_0^\infty f(x)\, dx$$

into an integral along a ray through the origin in the complex plane. As an illustrative example he takes

$$f(y) = x^{n-1} e^{-y},$$

and comes out with the results that Euler [1781] had obtained by the imaginary substitution $x = (p + qi)y$ (Section 1.9 above). We recall that

Legendre [1814a] also used this substitution (Section 1.17 above). Cauchy then gives some other examples.

In Section IV Cauchy takes

$$y = M + Ni = x(\cos z + i \sin z).$$

In the language of Section 2.4 above, he is transforming to polar coordinates. He takes $x_0 = z_0 = 0$; the curvilinear quadrilateral thus collapses to a sector in the (u, v)-plane with its vertex at the origin and bounded by the radial segments

$$v = 0, \quad v = u \tan Z \quad (0 \leqslant u \leqslant X)$$

and an arc of the circle $u^2 + v^2 = X^2$. He writes down the basic equations for this case, and gives one illustrative example, which is not of interest for the general theory.

This transformation, as we shall see in later chapters, was to become much more important in Cauchy's work than it is in the 1814 memoir.

In Section V Cauchy takes

$$M = ax^2, \quad N = xz.$$

In our geometrical interpretation (under the assumption $a > 0$) the coordinate curves are $u = $ const. (straight lines parallel to the v-axis in the right half-plane) and $v^2/u = $ const. (parabolas with vertex at the origin and axes parallel to the u-axis). He goes through the same routine as before, and gives one illustrative example, in which he takes $f(y) = e^{-y^2}$. Nothing of general interest emerges.

2.7. The final section (VI) of Part I is entitled 'De la séparation des exponentielles'. Since Cauchy is scrupulously avoiding complex integrands, he cannot use the now familiar device of replacing an integral of either of the forms

$$\int f(x) \cos ax \, dx, \quad \int f(x) \sin ax \, dx;$$

where $a > 0$, by

$$\int f(x) e^{iax} \, dx,$$

so that $e^{iay} = e^{ia(x+zi)}$ is small in absolute value when z is large and positive. In the present section he explains how he circumvents this difficulty.

He starts from a function of the form

$$p(y) = q(y) \cos r(y),$$

and follows the general pattern of Section I (Section 2.3 above), taking

$$y = M(x, z) + iN(x, z).$$

He expresses p, q and r as functions of (x, z) in equations of the form[3]

$$p(y) = P'(x, z) + iP''(x, z),$$
$$q(y) = Q'(x, z) + iQ''(x, z),$$
$$r(y) = R'(x, z) + iR''(x, z),$$

so that

$$P' + P''i = (Q' + Q''i)\cos(R' + R''i).$$

Writing, as in Section I,

$$S = P'\frac{\partial M}{\partial x} - P''\frac{\partial N}{\partial x},$$

with similar expressions for T, U, V, he expresses S, T, U, V in terms of Q', Q'', R', R'' and the partial derivatives of M and N, obtaining equations of the form

$$2S = S_1 e^{R''} + S_2 e^{-R''},$$

etc., where S_1 and S_2 do not involve R''. His next move is to claim that when the basic equations

$$\frac{\partial S}{\partial z} = \frac{\partial U}{\partial x}, \quad \frac{\partial T}{\partial z} = \frac{\partial V}{\partial x}$$

are expressed in this way, one is entitled to equate the coefficients of $e^{R''}$, and similarly those of $e^{-R''}$. Thus we must have, for instance,

$$\frac{\partial}{\partial z}(S_2 e^{-R''}) = \frac{\partial}{\partial x}(U_2 e^{-R''}),$$

$$\frac{\partial}{\partial z}(T_2 e^{-R''}) = \frac{\partial}{\partial x}(V_2 e^{-R''}).$$

He remarks that these equations can be verified directly. We note his tacit assumption that not only $p(y)$, but also $q(y)$ and $r(y)$ separately, possess derivatives for complex y.

Integration with respect to x and z over the intervals $(0, X)$ and $(0, Z)$ then gives, for instance,

$$\int_0^X S_2(x, Z) e^{-R''(x, Z)} \, dx - \int_0^X S_2(x, 0) e^{-R''(x,0)} \, dx = \int_0^Z U_2(X, z) e^{-R''(X,z)} \, dz$$

$$- \int_0^Z U_2(0, z) e^{-R''(0,z)} \, dz,$$

$$(2.7)$$

<hr/>

[3] In Section 2.3 we wrote $P + iQ$ for Cauchy's $P' + iP''$; in this section it is more convenient to retain Cauchy's notation, although we do modify his notation in some other respects, in the interests of clarity.

and a similar equation involving T_2 and V_2. Another pair of equations comes from the coefficients of $e^{R''}$.

With a view to later applications, he examines what equations (2.7) become in the most important special case, where $M = x$ and $N = z$. It turns out that we then have

$$S_2 = V_2 = Q' \cos R' - Q'' \sin R',$$

$$T_2 = -U_2 = Q'' \cos R' + Q' \sin R'.$$

Because of his usual assumption that $p(y)$ is real for real y, we have $Q''(x, 0) = 0$. Equation (2.7) thus reduces to

$$\int_0^X [Q'(x, Z) \cos R'(x, Z) - Q''(x, Z) \sin R'(x, Z)] e^{-R''(x, Z)} dx$$

$$- \int_0^X Q'(x, 0) \cos R'(x, 0) dx$$

$$= -\int_0^Z [Q''(X, z) \cos R'(x, Z) + Q'(X, z) \sin R'(X, z)] e^{-R''(X, z)} dz$$

$$+ \int_0^Z [Q''(0, z) \cos R'(0, z) + Q'(0, z) \sin R'(0, z)] e^{-R''(0, z)} dz.$$

Specialising these results in various ways, Cauchy examines a number of illustrative examples; we mention only a couple of them.

Taking $q(y) = 1$ and supposing that $e^{-R''(x,z)}$ vanishes when $z = +\infty$, he obtains

$$\int_0^X \cos R(x, 0) dx = \int_0^\infty e^{-R''(X,z)} \sin R'(X, z) dz$$

$$- \int_0^\infty e^{-R''(0,z)} \sin R'(0, z) dz;$$

his comment here is that the formula expresses an indefinite integral as a definite integral.

Another sample formula he gives is

$$\int_0^X \frac{\cos ax}{1 + x} dx = \int_0^\infty \frac{z \, e^{-az}}{1 + z^2} dz + \sin aX \int_0^\infty \frac{1 + X}{(1 + X)^2 + z^2} e^{-az} dz$$

$$- \cos aX \int_0^\infty \frac{z}{(1 + X)^2 + z^2} e^{-az} dz.$$

On this he remarks that, since all the integrands on the right-hand side are positive, such an equation might be useful for deriving an asymptotic expression for the left-hand side when X is large. Taking $X = +\infty$, he obtains the simple result

$$\int_0^\infty \frac{\cos ax}{1+x}\, dx = \int_0^\infty \frac{z\, e^{-az}}{1+z^2}\, dz.$$

We note for future reference that he is assuming here that if a function $h(X, z)$ vanishes when $X = +\infty$, the same holds for $\int_0^\infty h(X, z)\, dz$; this is an even stronger assumption than the one we remarked on in Section 2.5 above. Although this causes no trouble here, such assumptions did lead him into difficulties later (Sections 3.9 and 4.14 below).

Further and more interesting applications of the technique of separation of exponentials appear in Part II of the 1814 memoir (Section 2.16 below). However, the device lost its *raison d'être* when Cauchy began to admit complex-valued integrands, as he was to do within a few years (Section 3.3 below).

2.8. We now come to Part II of the 1814 memoir; it bears the general title 'Sur les difficultés que peut offrir l'intégration des équations différentielles'. Here Cauchy is not referring to general differential equations; his concern is with the integration of functions possessing singularities of some kind, usually infinities.

The main results of Part II are of a much more novel character than those of Part I, and opened the way to formulae of a completely new type; Cauchy reaches conclusions that are, in modern terms, equivalent to special cases of the residue theorem, which he himself did not state in a general form until 1831 (Section 6.15 below). He is led to these results by considering double integrals of functions having singularities in the domains of integration, and he uses them in ways that are familiar to us today for the evaulation of definite integrals of various types.

In Section III of this Part, however, he is concerned with the relationship between indefinite and definite integrals of a single real variable; this question raised some important issues of principle, which led in particular to Cauchy's definition of principal-value integrals [1822b]. These considerations were also relevant to his definition of continuity in the *Analyse algébrique* [1821a] and to his definition of the definite integral as a limit of approximating sums in the *Calcul infinitésimal* [1823b]. It will be convenient to discuss this section first, before examining his investigation of double integrals.

2.9. In Section III of Part II Cauchy is concerned with the relation between the definite integral

$$\int_a^b \varphi'(z)\, dz$$

and the primitive function $\varphi(z)$ of $\varphi'(z)$ in cases where $\varphi'(z)$ has an infinity in the interval (a, b). We recall that at this period the definite integral was usually defined to be the difference between the values of the primitive function at the two ends of the interval: that it was normally equal in some sense to the sum of the infinitesimals $\varphi'(z)\, dz$ was regarded as a theorem that had to be proved.

Some difficulties with this identification had already been encountered. These probably go back to d'Alembert [1768a], who remarks on the paradox that arises when one forms the definite integral of x^m for $m < -1$ over an interval including $x = 0$. Lacroix, in his *Traité* [1798, pp. 166–8], attributes the paradox to the inability of the analysis to express the geometrical situation, and suggests that the analytic treatment should be preferred: the same passage appears in the second edition [1814, pp. 161–3]. In his lectures at the Ecole Polytechnique, Lagrange gives the following example.[4] Let

$$y = f(x) = \frac{1}{a - x} - \frac{1}{a},$$

where $a > 0$. Then $f(0) = 0$, and

$$f'(x) = \frac{1}{(a - x)^2} > 0;$$

on the other hand, $f(x) < 0$ for $x > a$. As he remarks, this contradicts the idea that $f(x)$ is the sum of the infinitesimals $f'(x)\, dx$, which are all positive; he concludes that in this case the principles of the differential calculus are defective.

Cauchy begins his discussion by remarking that if $\varphi(z)$ increases or decreases in a continuous manner between $z = b'$ and $z = b''$, then the value of the definite integral will be $\varphi(b'') - \varphi(b')$, in conformity with the usual rule. The phrase 'in a continuous manner' used here does not refer to the then customary definition of continuity, where a function was said to be continuous if it was given by the same analytic expression throughout its domain of definition; it is rather an adumbration of Cauchy's later definition of continuity [1821a, Chapter II, Section 2]. He then says that if, on the other hand, $\varphi(z)$ changes its value abruptly as z passes through some value Z in the range of integration, then the usual procedure may give a wrong answer; to obtain the correct answer, one should proceed as follows. If ζ is small (and positive), then we should have

[4] Lagrange [1804], p. 69. *Oeuvres*, **10**, 90.

$$\int_{b'}^{b''} \varphi'(z)\,dz = \left[\int_{b'}^{Z-\zeta} \varphi'(z)\,dz + \int_{Z+\zeta}^{b''} \varphi'(z)\,dz\right]_{\zeta=0}$$

$$= [\varphi(b'') - \varphi(Z+\zeta) + \varphi(Z-\zeta) - \varphi(b')]_{\zeta=0}$$

$$= \varphi(b'') - \varphi(b') - \Delta,$$

where $\Delta = [\varphi(Z+\zeta) - \varphi(Z-\zeta)]_{\zeta=0}$.

Today we should interpret his expression for the correction term Δ as

$$\lim_{\zeta\to 0+} [\varphi(Z+\zeta) - \varphi(Z-\zeta)].$$

What he seems to be saying here amounts essentially to insisting that (in his own later terminology) a definite integral should be a continuous function of the upper limit of integration.

He illustrates his argument by considering

$$\int_{-2}^{4} \frac{dz}{z};$$

here $\varphi(z) = \log z$, which changes continuously except at $z = 0$, where we have

$$[\log \zeta - \log(-\zeta)]_{\zeta=0} = -\log(-1).$$

Thus $\Delta = -\log(-1)$, and he concludes that

$$\int_{-2}^{4} \frac{dz}{z} = \log 4 - \log(-2) + \log(-1) = \log 4 - \log 2 = \log 2.$$

He remarks that with this interpretation the integral from -2 to 4 is the same as that from 2 to 4; it follows that the integral from -2 to 2 vanishes, as it should, since the elements whose sum makes up the integral are equal in value and opposite in sign.

He also gives a second (rather artificial) example, mentioned in Smithies [1986] p. 53.

In the above argument we can see a foreshadowing of Cauchy's later formal definition [1822b] of the principal value of an integral in cases where the integrand has an infinity in the range of integration (Section 3.8 below). In fact, although Cauchy provides no formal definition here, he does investigate a number of principal-value integrals later in the 1814 memoir.

Several attempts were made to rehabilitate the old definition in cases where the integrand has an infinite singularity. One was made by Poisson [1820], and had important repercussions (Section 4.2 below); a very different one was made by M. V. Ostrogradskiĭ in an unpublished paper written in 1824 and has been discussed by Yushkevich [1965].

We also note that the idea of a correction term, analogous to the term Δ introduced above, was to play an important role elsewhere in the 1814 memoir, and contained the germ of the idea of residues.

2.10. In Section I of Part II Cauchy opens the discussion of double integrals in which the integrand has a singularity. He begins by recalling the results of Part I, Section I (Section 2.3 above); he writes down the equation[5]

$$\int_0^a \int_0^b \frac{\partial S}{\partial z}\, dx\, dz = \int_0^a \int_0^b \frac{\partial U}{\partial x}\, dx\, dz. \tag{2.8}$$

Here he insists that the double integrals must be regarded as being the sums of the elements $(\partial S/\partial z)\, dx\, dz$ and $(\partial U/\partial x)\, dx\, dz$ over the range of integration; if $\partial S/\partial z$ and $\partial U/\partial x$ both have determinate values for all such (x, z), then the double integrals will also have determinate values. In this case they can be evaluated as repeated integrals, and we obtain

$$\int_0^a S(x, b)\, dx - \int_0^a S(x, 0)\, dx = \int_0^b U(a, z)\, dz - \int_0^b U(0, z)\, dz, \tag{2.9}$$

as in Part I.

He now supposes that the function

$$\frac{\partial S}{\partial z} = \frac{\partial U}{\partial x}$$

'becomes indeterminate' when, say, $x = X$ and $z = Z$; in this case the double integral will be indeterminate, and equation (2.8) becomes meaningless.[6] In general, however, he says, the two repeated integrals will still exist, but we can no longer assume that they will be equal; equation (2.9) must therefore be replaced by an equation of the form

$$\int_0^a S(x, b)\, dx - \int_0^a S(x, 0)\, dx + A = \int_0^b U(a, z)\, dz - \int_0^b U(0, z)\, dz, \tag{2.10}$$

where A is a correction term. The problem of evaluating A is deferred to a later section.

He next examines how singularities can arise in practice. He writes (as in Part I, Section I)

---

[5] In the notation we used in Section 2.3, he is taking $x_0 = z_0 = 0$, $X = a$, $Z = b$.

[6] In his abstract of the 1814 memoir, Poisson [1814] remarks that, for this to happen, the integrand must have an infinity when $x = X$ and $z = Z$. In practice, as we shall see in Section 2.11 below, the integrand usually does take the form $0/0$ at (X, Z), and therefore appears indeterminate, but at the same time it is unbounded in the neighbourhood of (X, Z); in this sense, Cauchy and Poisson were both right.

$$f(M + Ni) = P' + P''i,$$

where M and N one supposed to be well-behaved functions, and derives the equations

$$S = P' \frac{\partial M}{\partial x} - P'' \frac{\partial N}{\partial x}, \quad U = P' \frac{\partial M}{\partial z} - P'' \frac{\partial N}{\partial z};$$

hence, he says, singularities can appear only when P' and P'' can be expressed as fractions with denominators having zeros. To bring this out, he supposes that[7]

$$f(y) = \frac{G(y)}{F(y)},$$

where, say, $G(M \pm Ni) = Q' \pm Q''i$, $\quad F(M \pm Ni) = R' \pm R''i$; from these equations it follows that

$$P' = \frac{Q'R' + Q''R''}{R'^2 + R''^2}; \quad P'' = \frac{Q''R' - Q'R''}{R'^2 + R''^2}.$$

Tacitly assuming that Q' and Q'' are well-behaved, he concludes that for P' or P'' to have a singularity we must have $R' = R'' = 0$. He points out that there is no need to eliminate z or x from the equations $R' = R'' = 0$, since we have

$$R'^2 + R''^2 = F(M + Ni)F(M - Ni).$$

In order to determine the position of the singularity, one only has to find the roots $y = \alpha + \beta i$ of the equation $F(y) = 0$, and then solve the equations $M = \alpha$, $N = \beta$.

The remainder of Section I is devoted to illustrative examples. Thus, in the case $M = x$, $N = z$, he takes for $F(y)$ such functions as $y^{2n} + 1$, $e^y - 1$ and $a - \cos 2y$ (both for $0 \leqslant a < 1$ and for $a > 1$).

In the case $M = ax$, $N = xz$ (Section 2.6 above), so that $S = aP' - zP''$, he remarks that S can be indeterminate for $z = \infty$, $P'' = 0$. This leads him to some curious conclusions; for example, with $F(y) = e^y - 1$, giving $\alpha + \beta i = 2k\pi i$, he gets

$$X = 0, \quad Z = 2k\pi/0.$$

2.11. Section II of Part II is devoted to a general discussion of the evaluation of the correction term A appearing in equation (2.10) and in similar equations. Cauchy begins by rephrasing the problem as follows: suppose that the function $\varphi(x, z)$ becomes indeterminate when $x = X$ and

---

[7] We have written $G(y)$ for Cauchy's $\mathcal{F}(y)$.

$z = Z$, and that we evaluate the integral

$$\int_{a'}^{a''}\int_{b'}^{b''} \frac{\partial \varphi}{\partial z}\, dx\, dz,$$

where $a' \leqslant X \leqslant a''$, $b' \leqslant Z \leqslant b''$, as a repeated integral, with both orders of integration.

For simplicity, he begins with the case $X = a'$, $Z = b'$, and assumes that $\varphi(x, z)$ is well-behaved elsewhere in the range of integration. Performing the z-integration first, one gets just

$$\int_{a'}^{a''} [\varphi(x, b'') - \varphi(x, b')]\, dx.$$

He then considers what happens if the x-integration is performed first. Since $\varphi(x, z)$ is well-behaved for $a' \leqslant x \leqslant a''$, $b' + \zeta \leqslant z \leqslant b''$, where ζ is small and positive, the double integral over this range can be evaluated as a repeated integral in either order, so that we get

$$\int_{b'+\zeta}^{b''} dz \int_{a'}^{a''} \frac{\partial \varphi}{\partial z}\, dx = \int_{a'}^{a''} [\varphi(x, b'') - \varphi(x, b' + \zeta)]\, dx. \qquad (2.11)$$

To obtain the correct value of

$$\int_{b'}^{b''} dz \int_{a'}^{a''} \frac{\partial \varphi}{\partial z}\, dx,$$

we must therefore put $\zeta = 0$ in the right-hand side of (2.11).

By the definition of the correction term A, this result must be equal to

$$\int_{a'}^{a''} [\varphi(x, b'') - \varphi(x, b')]\, dx + A.$$

The term involving $\varphi(x, b'')$ can now be cancelled from both sides of the equation, and we are left with

$$A = \left[\int_{a'}^{a''} (\varphi(x, b') - \varphi(x, b' + \zeta))\, dx\right]_{\zeta=0}. \qquad (2.12)$$

From a modern point of view, we should more naturally write this equation as

$$A = \lim_{\zeta \to 0+} \int_{a'}^{a''} [\varphi(x, b') - \varphi(x, b' + \zeta)]\, dx.$$

Cauchy then makes a further reduction, pointing out that if ε is small and positive, the values of x with $a' + \varepsilon \leqslant x \leqslant a''$ make no contribution to the expression in equation (2.12), so that we can write

$$A = \left[\int_0^\varepsilon (\varphi(a' + \xi, b') - \varphi(a' + \xi, b' + \zeta))\,d\xi\right]_{\zeta=0}.$$

He goes on to say that, if ε is small enough, the term

$$\int_0^\varepsilon \varphi(a' + \xi, b')\,d\xi$$

will be negligible, so that we shall have finally

$$A = -\left[\int_0^\varepsilon \varphi(a' + \xi, b' + \zeta)\,d\xi\right]_{\zeta=0}.$$

From a modern point of view, what Cauchy is rather clumsily trying to say can be written as

$$A = -\lim_{\varepsilon \to 0+} \lim_{\zeta \to 0+} \int_0^\varepsilon \varphi(a' + \xi, b' + \zeta)\,d\xi.$$

The lack of an efficient notation for indicating the order in which repeated limits are to be taken was to prove a stumbling block for Cauchy on a number of occasions; however, all is well here.

At this point Cauchy tabulates the results for this and similar cases as follows.

$$\left.\begin{aligned}
&\text{if } X = a',\ Z = b',\ \text{then } A = -\int_0^\varepsilon \varphi(X + \xi,\ Z + \xi)\,d\xi;\\[2mm]
&\text{if } X = a',\ Z = b'',\ \text{then } A = +\int_0^\varepsilon \varphi(X + \xi,\ Z - \xi)\,d\xi;\\[2mm]
&\text{if } X = a'',\ Z = b',\ \text{then } A = -\int_0^\varepsilon \varphi(X - \xi,\ Z + \zeta)\,d\xi;\\[2mm]
&\text{if } X = a'',\ Z = b'',\ \text{then } A = +\int_0^\varepsilon \varphi(X - \xi,\ Z - \zeta)\,d\xi.
\end{aligned}\right\} \quad (2.13)$$

In each case, the integral has to be interpreted in the way described above. He here introduces the term *singular integral* for expressions arising in this manner, i.e. non-zero definite integrals taken over an infinitely small interval.

As an illustrative example, he takes

$$\varphi(x, z) = \frac{z}{x^2 + z^2}$$

over the domain $0 \leqslant x \leqslant 1$, $0 \leqslant z \leqslant 1$. Here $X = Z = 0$. Integrating $\partial\varphi/\partial z$ with respect to z first, we get

$$\int_0^1 \frac{dx}{x^2 + 1} = \frac{\pi}{4}$$

for the value of the repeated integral. The correction term should therefore be

$$A = - \left[\int_0^\varepsilon \frac{\zeta}{\xi^2 + \zeta^2} \, d\xi \right]_{\zeta=0}$$

$$= - \left[\arctan \frac{\varepsilon}{\zeta} \right]_{\zeta=0}$$

$$= - \frac{\pi}{2}.$$

The result of performing the x-integration first should therefore be $(\pi/4) - (\pi/2) = -(\pi/4)$, and Cauchy verifies directly that this is correct. We note that this example is closely related to the one given by d'Alembert [1768b] and described in Section 1.3 above.

Cauchy now considers the case where $a' < X < a''$, $b' < Z < b''$, so that the singularity is in a general position: combining the earlier results, he obtains

$$A = \int_0^\varepsilon [\varphi(X - \xi, z - \zeta) + \varphi(X + \xi, Z - \zeta)$$

$$- \varphi(X - \xi, Z + \xi) - \varphi(X + \xi, Z + \xi)] \, d\xi, \qquad (2.14)$$

to be interpreted as before.

To illustrate this case, he takes

$$\varphi(x, z) = \frac{1 + x^2 - z^2}{(1 + x^2 - z^2)^2 + 4x^2 z^2},$$

the domain of integration being $-a \leqslant x \leqslant a$, $0 \leqslant z \leqslant b$, where $b > 1$. We note that, if we write $y = x + zi$, then $\varphi(x, z)$ is the real part of

$$f(y) = \frac{1}{1 + y^2}.$$

Here $X = 0$, $Z = 1$, so that

$$\varphi(X + \xi, Z + \zeta) = \frac{\xi^2 - 2\zeta - \zeta^2}{(\xi^2 - 2\zeta - \zeta^2)^2 + 4\xi^2(1 + \zeta)^2}.$$

Although in this case the integrations required to evaluate the correction term from equation (2.14) above can be carried out directly. Cauchy prefers, with a view to his next application, to approximate by neglecting higher powers of ξ and ζ; he justifies his approximations with some care, finally replacing $\varphi(X + \xi, Z + \zeta)$ by

$$\frac{\xi^2 - 2\zeta}{4(\xi^2 + \zeta^2)}.$$

From this expression he then derives the result $A = \pi$.

He then modifies this example by taking

$$\varphi(x, z) = \frac{1 + x^2 - z^2}{(1 + x^2 - z^2)^2 + 4x^2z^2} \psi(x, z).$$

where $\psi(x, z)$ 'does not become indeterminate' in the range of integration, which is the same as before. Using the same approximation technique, he concludes that in this case $A = \pi\psi(0, 1)$.

He points out that if either X or Z coincides with one of the limits of integration, then equation (2.14) will contain only two of the terms listed in (2.13) above. Also, if $\varphi(x, z)$ becomes indeterminate for several values of (X, Z) in the domain of integration, then the correction term A will be the sum of their separate contributions.

In some general remarks at the end of Section II, he says that he will not be discussing cases where $\varphi(x, z)$ itself becomes infinite when $x = X$, $z = Z$, since this does not usually happen in the theory he wishes to develop; he seems to mean that when one substitutes these values in $\varphi(x, z)$ it should appear in the form $0/0$. As is clear from the examples described above, this does not exclude the possibility that $\varphi(x, z)$ may be unbounded in the neighbourhood of the singularity; see also our remarks on Poisson's abstract in Section 2.10 above.

In this section Cauchy has succeeded in finding a technique for evaluating the correction term A without explicitly carrying out the two repeated integrations. It is perhaps worth noting that at this stage his results do not yet require that $\varphi(x, z)$ should be the real part of an analytic function, as we can see from the following example. Take

$$\varphi(x, z) = \frac{z}{x^2 + 2z^2};$$

then it is easily verified that

$$\int_0^1 dx \int_0^1 \frac{\partial\varphi}{\partial z}\, dz = \frac{1}{\sqrt{2}} \arctan(1/\sqrt{2}),$$

$$\int_0^1 dz \int_0^1 \frac{\partial\varphi}{\partial z}\, dx = -\frac{1}{\sqrt{2}} \arctan(\sqrt{2}),$$

whence

$$A = -\frac{1}{\sqrt{2}} [\arctan(1/\sqrt{2}) + \arctan(\sqrt{2})]$$

$$= -\pi/2\sqrt{2}.$$

We also have

$$-\lim_{\zeta\to 0+}\int_0^\varepsilon \varphi(\xi,\,\zeta)\,d\xi = -\lim_{\zeta\to 0+}\frac{1}{\sqrt{2}}\arctan\left(\varepsilon/\zeta\sqrt{2}\right)$$

$$= -\pi/2\sqrt{2} = A,$$

as it should be.

In cases where $\varphi(x,\,z)$ does happen to be the real or imaginary part of an analytic function $f(x+zi)$, what Cauchy has done is equivalent to evaluating the real or imaginary part of the residue of $f(y) = f(x+zi)$ at an isolated singularity $y = X + Zi$, as we shall see in our discussion of Section IV (Section 2.12 below).

2.12. In Section IV of Part II, Cauchy goes more deeply into the problem of evaluating the correction term. Whereas in Section II he does not specify in detail the nature of the singularity he is dealing with, in Section IV, as we shall see, he requires the singularity to be (in modern language) a simple pole.

He begins by summarizing the results of Section II, quoting the basic equation (2.14) above. He then returns to the situation described in Section I (Section 2.10 above), writing, as before,

$$f(M \pm Ni) = P' \pm P''i,$$

$$S = P'\frac{\partial M}{\partial x} - P''\frac{\partial N}{\partial x},$$

$$T = P'\frac{\partial N}{\partial x} + P''\frac{\partial M}{\partial x}.$$

The correction term for S will then be

$$A = \int_{b'}^{b''} dz \int_{a'}^{a''} \frac{\partial S}{\partial z}\,dx - \int_{a'}^{a''} dx \int_{b'}^{b''} \frac{\partial S}{\partial z}\,dz.$$

As in Section I, he supposes that

$$f(y) = \frac{G(y)}{F(y)},$$

where $G(y)$ is tacitly assumed to be a well-behaved function, and $F(y)$ has a zero at $y = \alpha + \beta i$, which he expresses in the form

$$\alpha + \beta i = M(X,\,Z) + N(X,\,Z)i,$$

where $a' \leqslant X \leqslant a''$, $b' \leqslant Z \leqslant b''$.

Under his usual tacit assumption that $f(y)$ is real for real y. Cauchy obtains the equations

$$P' = \frac{1}{2}\left[\frac{G(M+Ni)}{F(M+Ni)} + \frac{G(M-Ni)}{F(M-Ni)}\right],$$

$$P'' = \frac{1}{2i}\left[\frac{G(M+Ni)}{F(M+Ni)} - \frac{G(M-Ni)}{F(M-Ni)}\right].$$

Denoting the values assumed by $\partial M/\partial x$, $\partial N/\partial x$, $\partial M/\partial z$, $\partial N/\partial z$, when $x = X$, $z = Z$ by $\partial M/\partial X$, $\partial N/\partial X$, $\partial M/\partial Z$, $\partial N/\partial Z$, respectively, he then introduces the notation

$$B = \left(\frac{\partial M}{\partial X}\right)^2 + \left(\frac{\partial N}{\partial X}\right)^2,$$

$$C = \frac{\partial M}{\partial X}\cdot\frac{\partial M}{\partial Z} + \frac{\partial N}{\partial X}\cdot\frac{\partial N}{\partial Z},$$

$$D = \left(\frac{\partial M}{\partial Z}\right)^2 + \left(\frac{\partial N}{\partial Z}\right)^2,$$

$$E = \frac{\partial M}{\partial X}\cdot\frac{\partial N}{\partial Z} - \frac{\partial M}{\partial Z}\cdot\frac{\partial N}{\partial X}.$$

We note that E is the Jacobian of the transformation from the (x, z)-plane to the (M, N)-plane, and that $E^2 = BD - C^2$.

Tacitly assuming for the moment (although he makes the assumption explicit later, as we shall see) that $\alpha + \beta i$ is a simple zero of $F(y)$, he also writes

$$\left.\begin{array}{l} \lambda = \dfrac{1}{2}\left[\dfrac{G(\alpha - \beta i)}{F'(\alpha - \beta i)} + \dfrac{G(\alpha + \beta i)}{F'(\alpha + \beta i)}\right], \\[3mm] \mu = \dfrac{1}{2i}\left[\dfrac{G(\alpha - \beta i)}{F'(\alpha - \beta i)} - \dfrac{G(\alpha + \beta i)}{F'(\alpha + \beta i)}\right]. \end{array}\right\} \tag{2.15}$$

He then takes

$$x = X + \xi,\ z = Z + \zeta,$$

where ξ and ζ are assumed to be small, and he introduces the following approximations:

$$M(X + \xi, Z + \zeta) = \alpha + \frac{\partial M}{\partial X}\xi + \frac{\partial N}{\partial Z}\zeta,$$

$$N(X + \xi, Z + \zeta) = \beta + \frac{\partial N}{\partial X}\xi + \frac{\partial N}{\partial Z}\zeta,$$

$$F(M \pm Ni) = \left[\left(\frac{\partial M}{\partial X}\xi + \frac{\partial M}{\partial Z}\zeta\right) \pm \left(\frac{\partial N}{\partial X}\xi + \frac{\partial N}{\partial Z}\zeta\right)i\right]F'(\alpha + \beta i),$$

$$G(M \pm Ni) = G(\alpha \pm \beta i).$$

Without making any further approximations, he arrives at the expression

$$S(X + \xi, Z + \zeta) = \frac{\lambda(B\xi + C\zeta) - \mu E\zeta}{B\xi^2 + 2C\xi\zeta + D\zeta^2}.$$

At this stage he assumes that B/E is positive; since $B \geq 0$ in any case, this amounts to saying that $B > 0$ and $E > 0$.

To obtain the correction term for S, he then has to evaluate the expression

$$\left[\int_0^\varepsilon S(X + \xi, Z + \zeta) \, d\xi \right]_{\xi=0}$$

for small values of ε, together with three similar expressions, as in (2.13) above. He finds

$$\int_0^\varepsilon S(X + \xi, Z + \zeta) \, d\xi = \frac{1}{2} \lambda \log \left(\frac{B\varepsilon^2 + 2C\varepsilon\zeta + D\zeta^2}{D\zeta^2} \right)$$

$$- \mu \left(\arctan \frac{B\varepsilon + C\xi}{E\zeta} - \arctan \frac{C}{E} \right),$$

and three similar expressions. Putting the four results together, as in equation (2.14) of Section 2.11, gives

$$\lambda \log \left(\frac{B\varepsilon^2 - 2C\varepsilon\zeta + D\zeta^2}{B\varepsilon^2 + 2C\varepsilon\zeta + D\zeta^2} \right) + 2\mu \arctan \frac{B\varepsilon + C\zeta}{E\zeta} + 2\mu \arctan \frac{B\varepsilon - C\zeta}{E\zeta}.$$

He notes that when ζ becomes 0 the logarithmic term disappears, and the other two terms contribute

$$2\mu \left(\frac{\pi}{2} + \frac{\pi}{2} \right) = 2\mu\pi.$$

He also examines the cases where only two or only one of the terms may be involved. He tabulates his results as follows:

(i) if $a' < X < a''$, $b' < Z < b''$, then $A = 2\mu\pi$;
(ii) if just one of X and Z takes an extreme value, then $A = \mu\pi$;
(iii) if both X and Z take extreme values, then one obtains an expression (in general meaningless), which he writes as

$$A = \mu \left(\frac{\pi}{2} \pm \arctan \frac{C}{E} \right) \pm \infty \cdot \lambda.$$

If $\lambda = 0$ in this case, however, one obtains a finite value for A; if (X, Z) is (a', b') or (a'', b''), then

$$A = \mu \left(\frac{\pi}{2} - \arctan \frac{C}{E} \right);$$

if (X, Z) is (a', b'') or (a'', b'), then

$$A = \mu \left(\frac{\pi}{2} + \arctan \frac{C}{E} \right).$$

Using the same argument for T as for S, he concludes that the correction term for T is $2\lambda\pi$ in case (i) above and $\lambda\pi$ in case (ii); in case (iii), the result is finite only if $\mu = 0$.

At this stage Cauchy remarks that his expressions for the correction terms become illusory if $F'(\alpha + \beta i)$ is zero or infinity, i.e. if $\alpha + \beta i$ is not a simple root of the equation $F(y) = 0$. He promises to examine the case of multiple roots later, but he does not do so in the 1814 memoir.

He also expresses some surprise at the fact that the values he has obtained for the correction terms depend only on the form of the function

$$f(y) = \frac{G(y)}{F(y)}$$

and on the roots of the equation $F(y) = 0$, but are independent of the choice of the functions $M(x, z)$ and $N(x, z)$; in particular the expressions he denoted by B, C, D, E have all disappeared. Indeed, he takes the trouble to verify directly that one gets the same answers in the special case $M = x$, $N = z$, as in the general case he has been discussing. It is clear that at this stage of his investigations he has not yet achieved a full understanding of what is going on.

From a modern point of view, and indeed in the light of Cauchy's own later work, what he has done is to show that the residue of $f(y) = G(y)/F(y)$ at a simple pole $y = \alpha + \beta i$ is given by

$$\lambda - \mu i = \frac{G(\alpha + \beta i)}{F'(\alpha + \beta i)}.$$

To some extent, he has further obscured the situation by expressing the real and imaginary parts of $G(\alpha + \beta i)/F'(\alpha + \beta i)$ in the special form given in equations (2.15).

In the remainder of Section IV Cauchy examines what modifications of his results are needed when one has to use the method of separation of exponentials, as in Part I, Section VI (Section 2.7 above). In the notation used there, one has to deal with double integrals such as

$$\iint \frac{\partial}{\partial z}(S_2 \, e^{-R''}) \, dx \, dz, \quad \iint \frac{\partial}{\partial z}(T_2 \, e^{-R''}) \, dx \, dz.$$

As he did earlier in Section IV, he supposes that $q(y)$ (which he now denotes by $f(y)$) is of the form

$$f(y) = \frac{G(y)}{F(y)},$$

where $G(y)$ is well behaved and $F(y)$ has a simple zero at $\alpha + \beta i$; he writes X and Z for values of x and z such that

$$M(x, z) = \alpha, \qquad N(x, z) = \beta$$

and (x, z) lies in the domain of integration. We shall not describe his arguments in detail; his main conclusion is that the correction terms for the above integrals can be obtained from those for the cases already considered by replacing λ and μ by $\gamma\lambda + \delta\mu$ and $\gamma\mu - \delta\lambda$, respectively, where

$$\gamma = e^{-R''(X, Z)} \cos R'(X, Z), \qquad \delta = e^{-R''(X, Z)} \sin R'(X, Z).$$

In other words, the new residue is

$$(\lambda - \mu i)(\gamma + \delta i) = (\lambda - \mu i) e^{i(R' + iR'')}.$$

2.13. The remaining sections of Part II are devoted to applications of the general results. We shall confine our attention to some passages that raise points of general interest.

The long Section V deals with cases where

$$M(x, z) = x, \qquad N(x, z) = z.$$

In the notation of Section 2.3 above, this implies that

$$S = V = P, \qquad T = -U = Q.$$

He writes a, b instead of X, Z, respectively. Equations (2.5) now have to be replaced by

$$\left.\begin{aligned}
\int_0^a P(x, b)\, dx - \int_0^a P(x, 0)\, dx + A' &= \int_0^b Q(0, z)\, dz - \int_0^b Q(a, z)\, dz, \\
\int_0^a Q(x, b)\, dx + A'' &= \int_0^b P(a, z)\, dz - \int_0^b P(0, z)\, dz,
\end{aligned}\right\}$$

$$(2.16)$$

where A' and A'' are correction terms.

As in Section 2.12, he supposes that

$$f(y) = \frac{G(y)}{F(y)},$$

where $G(y)$ is a well-behaved function. We recall that if $\alpha + \beta i$ is a simple zero of $F(y)$, then the corresponding values of λ and μ are given by

$$\lambda = \frac{1}{2}\left[\frac{G(\alpha - \beta i)}{F'(\alpha - \beta i)} + \frac{G(\alpha + \beta i)}{F'(\alpha + \beta i)}\right],$$

$$\mu = \frac{1}{2i}\left[\frac{G(\alpha - \beta i)}{F'(\alpha - \beta i)} - \frac{G(\alpha + \beta i)}{F'(\alpha + \beta i)}\right].$$

Since the contributions from all the zeros of $F(y) = F(x + zi)$ with $0 \leqslant x \leqslant a$, $0 \leqslant z \leqslant b$, have to be summed to obtain the correction terms

A' and A'', Cauchy introduces here some new notation; he writes $S(\mu_{\alpha,\beta})$ for the sum of the values of μ corresponding to zeros $\alpha + \beta i$ of $F(y)$ with $0 < \alpha < a$, $0 < \beta < b$, $S(\mu_{0,\beta})$ for the sum corresponding to zeros with $\alpha = 0$, $0 < \beta < b$, and $S(\lambda_{\alpha,\beta})$, $S(\lambda_{\alpha,0})$, $S(\lambda_{0,\beta})$ for similar sums over values of λ. Because of the assumptions that $F(y)$ and $G(y)$ are real for real y, the values of μ for zeros with $\beta = 0$ vanish automatically.

In this notation, the correction term for the double integral

$$\int_0^b \int_0^a \frac{\partial P}{\partial z} \, dx \, dz$$

is

$$A' = \pi[2S(\mu_{\alpha,\beta}) + S(\mu_{0,\beta})], \tag{2.17}$$

that for the double integral

$$\int_0^b \int_0^a \frac{\partial Q}{\partial z} \, dx \, dz$$

is

$$A'' = \pi[2S(\lambda_{\alpha,\beta}) + S(\lambda_{\alpha,0}) + S(\lambda_{0,\beta})], \tag{2.18}$$

provided that $F(0) \neq 0$. If $F(0) = 0$, this becomes

$$A'' = \pi[2S(\lambda_{\alpha,\beta}) + S(\lambda_{\alpha,0}) + S(\lambda_{0,\beta}) + \tfrac{1}{2}\lambda_{0,0}].$$

In applying these results to specific examples, Cauchy is interested almost entirely in cases where either a or b is infinite. He gives some general results for cases of this kind; a typical one is as follows[8] (we have made some trivial changes in notation).

Let

$$f(y) = \frac{G(y)}{F(y)},$$

the equation $1/f(y) = 0$ having only simple roots, and suppose that the real and imaginary parts of

$$\frac{G(x + zi)}{F(x + zi)}$$

are continuous[9] functions of z, vanishing when $x = \pm\infty$ and when $z = +\infty$. For each non-real root $\alpha + \beta i$ of $1/f(y) = 0$ write, as before,

$$\mu = \frac{1}{2i}\left[\frac{G(\alpha - \beta i)}{F'(\alpha - \beta i)} - \frac{G(\alpha + \beta i)}{F'(\alpha + \beta i)}\right].$$

[8] Cauchy [1814], p. 713; *Ouevres*, **(1)1**, 427.
[9] It is not clear whether the word 'continuous' is here being used in Euler's sense or in the more modern sense that Cauchy was to give it later in his *Analyse algébrique* [1821a]; the former meaning is perhaps the more likely one in this context.

Then

$$\int_{-\infty}^{\infty} f(x)\, dx = 2\pi S(\mu),$$

where $S(\mu)$ denotes the sum of the values of μ corresponding to roots $\alpha + \beta i$ with $\beta > 0$. He adds the remark that, if f is an even function, then

$$\int_{0}^{\infty} f(x)\, dx = \pi S(\mu).$$

He obtains this result by taking first $a = b = +\infty$ and then $a = -\infty$, $b = +\infty$, in the first equation (2.16), and using equation (2.17) for A'; in doing so, he is assuming that the terms

$$\int_{0}^{a} P(x,\, b)\, dx, \qquad \int_{0}^{b} Q(a,\, z)\, dz$$

will disappear automatically because $P(x,\, b)$ and $Q(a,\, z)$ vanish when $b = +\infty$ and $a = \pm\infty$, respectively. As we remarked in Section 2.7 above, this is an even stronger assumption than the one we commented on in Section 2.5 above.

We see that, apart from the inadequacy of his conditions at infinity, Cauchy has here obtained a version of one of the now standard applications of complex function theory to the evaluation of definite integrals over the range $(-\infty,\, +\infty)$.

2.14. His first specific example leads to the integral

$$\int_{0}^{\infty} \frac{x^{a-1}\, dx}{1 + x^{b}} = \frac{\pi}{b \sin(a\pi/b)} \qquad (0 < a < b).$$

He proves this first for the case where $a - 1$ and b are even integers, and then obtains it for general real a and b by appealing to the principle of the generality of analysis, a type of argument that he was later to reject.[10] The result was first given by Euler,[11] to whom Cauchy gives a reference. Euler's proofs apply when a and b are integers, but he also shows in his Section 369 how the result can be extended to cover the case where a and b are real numbers with a/b rational.

Cauchy uses a similar method to obtain the formula

$$\int_{0}^{\infty} \frac{x^{a-1}\, dx}{1 - x^{b}} = \frac{\pi}{b \tan(a\pi/b)} \qquad (0 < a < b),$$

where the integral is to be interpreted as what he was later to call a 'principal value'.

---

[10] Smithies [1986], p. 46.   [11] Euler [1768a], Sections 351 and 368.

After examining some further examples, Cauchy makes the general remark that his method can be used to evaluate integrals of the form

$$\int_{-\infty}^{\infty} \frac{P(x)}{Q(x)} \, dx,$$

where P and Q are real polynomials, provided that Q has only simple zeros; he points out that in this way one can avoid the task of decomposing the integrand into partial fractions. Only the non-real zeros of Q need to be taken into account, since its real zeros do not contribute to the result, if the integral is treated as a principal value.

He mentions a device used by Legendre [1814a, p. 126] to interpret such integrals, and illustrates it by splitting the integral

$$\int_0^{\infty} \frac{x^{-p}}{x^n - x^{-n}} \cdot \frac{dx}{x},$$

where $n > p$, into two parts, over $(0, 1)$ and $(1, \infty)$, respectively, and putting $x = 1/y$ in the second part, thus transforming the integral into

$$\int_0^1 \frac{x^{-p} - x^p}{x^n - x^{-n}} \cdot \frac{dx}{x}.$$

A similar device was used by Bidone [1812].

The next class of integrals that Cauchy considers are those of the form

$$\int_{-\infty}^{\infty} \frac{G(x) \, dx}{1 + x^2},$$

where $G(x)$ is well behaved, apart from $1/G(x)$ having simple zeros on the real axis; under his usual conditions at infinity, he concludes that

$$\int_{-\infty}^{\infty} \frac{G(x) \, dx}{1 + x^2} = \frac{\pi}{2}[G(i) + G(-i)].$$

In particular, if $G(x)$ is an even function, this reduces to

$$\int_0^{\infty} \frac{G(x) \, dx}{1 + x^2} = \frac{\pi}{2} G(i).$$

He uses this result to evaluate a group of principal-value integrals, these are (with $a < b$ throughout)

$$\int_0^{\infty} \frac{\sin ax}{\sin bx} \cdot \frac{dx}{1 + x^2} = \frac{\pi}{2} \cdot \frac{e^a - e^{-a}}{e^b - e^{-b}}, \tag{2.19}$$

$$\int_0^{\infty} \frac{\cos ax}{\cos bx} \cdot \frac{dx}{1 + x^2} = \frac{\pi}{2} \cdot \frac{e^a + e^{-a}}{e^b + e^{-b}}, \tag{2.20}$$

$$\int_0^{\infty} \frac{\sin ax}{x \cos bx} \cdot \frac{dx}{1 + x^2} = \frac{\pi}{2} \cdot \frac{e^a - e^{-a}}{e^b + e^{-b}}, \tag{2.21}$$

$$\int_0^\infty \frac{x\cos ax}{\sin bx} \cdot \frac{dx}{1+x^2} = \frac{\pi}{2} \cdot \frac{e^a + e^{-a}}{e^b - e^{-b}}. \tag{2.22}$$

These results were to play an important role later in the 1814 memoir (Section 2.17 below). Cauchy notes that by letting a become 0 in (2.20) and (2.22) one obtains

$$\int_0^\infty \frac{1}{\cos bx} \cdot \frac{dx}{1+x^2} = \frac{\pi}{e^b + e^{-b}},$$

$$\int_0^\infty \frac{x}{\sin bx} \cdot \frac{dx}{1+x^2} = \frac{\pi}{e^b - e^{-b}}.$$

The second of these was also given by Legendre [1814a, p. 125].

In another set of examples Cauchy examines what happens to the basic equations (2.16) when one takes $a = +\infty$, under the usual assumption that $P(x, z)$ and $Q(x, z)$ vanish for $x = +\infty$. The equations then reduce to

$$\left.\begin{aligned}
\int_0^\infty P(x, b)\,dx - \int_0^\infty P(x, 0)\,dx &= \int_0^b Q(0, z)\,dz - A', \\
\int_0^\infty Q(x, b)\,dx &= -\int_0^b P(0, z)\,dz - A''.
\end{aligned}\right\} \tag{2.23}$$

He applies (2.23) to the function

$$f(y) = \frac{y^m}{e^y - 1},$$

showing that A'', regarded as a function of b, changes abruptly whenever b passes through an integral multiple of 2π. This phenomenon occurs again later in a more important context (Sections 2.16 and 2.17 below).

Numerous other definite integrals are evaluated in Section V; although some of these results are interesting on their own account, they do not seem to raise any points of interest for general theory.

2.15. In Section VI of Part II Cauchy turns to the evaluation of the correction terms for the case discussed in Section III of Part I (Section 2.6 above). Here P and Q are defined by

$$P \pm Qi = f(ax \pm xzi),$$

and the formulae to be corrected are

$$a\int_0^c P(x, c)\,dx - b\int_0^c Q(x, c)\,dx - a\int_0^c P(x, 0)\,dx = -c\int_0^b Q(b, z)\,dz,$$

$$b\int_0^c P(x, c)\,dx + a\int_0^c Q(x, c)\,dx = c\int_0^b P(c, z)\,dz.$$

The correction terms A' and A'' to be added to the left-hand sides of these equations are given by the formulae (2.17) and (2.18), where the sums are now to be taken over the values of α and β for which $X = \alpha/a$ and $Z = a\beta/\alpha$ lie between the given limits. The only specific example he gives is one in which the correction terms vanish, so the details need not concern us.

In the 1814 memoir Cauchy does not consider the correction terms for the cases of Section III (polar coordinates) and Section IV (parabolic coordinates) of Part I. The polar-coordinate case was eventually to become of major importance, as we shall see in later chapters.

2.16. The main body of the memoir ends with Part II, Section VII. Here Cauchy is concerned with further developments of the results employing the separation of exponentials and discussed at the end of his Section IV (Section 2.12 above). He now confines his attention to the special case where $M = x$ and $N = z$, so that (in the notation of Part I, Section VI) we have

$$S_2 = V_2 = Q' \cos R' - Q'' \sin R',$$

$$T_2 = -U_2 = Q'' \cos R' + Q' \sin R'.$$

There is no need to discuss the general results of this section in detail, but some of the illustrative examples had important repercussions, as we shall see in Section 2.17. In Section V of Part II (Section 2.14 above) Cauchy had shown, without using the separation of exponentials, that

$$\int_0^\infty \frac{\sin ax}{\sin bx} \cdot \frac{dx}{1 + x^2} = \frac{\pi}{2} \cdot \frac{e^a - e^{-a}}{e^b - e^{-b}} \qquad (-b < a < b). \tag{2.24}$$

By using the results at the end of Section IV, he obtains the expression

$$\frac{\pi}{2b} - \frac{\pi e^{-a}}{e^b - e^{-b}} + \frac{\pi}{b} \sum_{n=1}^\infty \frac{(-1)^n \cos{(n\pi a/b)}}{1 + (n^2\pi^2/b^2)} \tag{2.25}$$

for this integral, where a and b are now arbitrary real numbers with $b \neq 0$. He observes that the infinite series on the right-hand side of (2.25) represents a periodic function of a with period $2b$; using the known value (2.24) for $-b < a < b$, he obtains the expression

$$\frac{\pi}{2} \cdot \frac{e^{rb} + e^{-rb} - 2e^{-a}}{e^b - e^{-b}}$$

for the integral, where r denotes the absolute value of the difference between a/b and the nearest even integer. We note that in this case the result is a continuous function of a.

With a view to the discussion in the next section, we also quote here the formula

$$\int_0^\infty \frac{x\cos ax}{\sin bx} \cdot \frac{dx}{1+x^2} = \frac{\pi}{2} \cdot \frac{e^{br} - e^{-br} + 2e^{-a}}{e^b - e^{-b}},$$

where r now denotes the difference (not its absolute value this time) between a/b and the nearest even integer. It turns out that the value of this integral is discontinuous as a function of a at every odd multiple of b. For instance, its limit as $a \to b-$ is

$$\frac{\pi}{2} \cdot \frac{e^b + e^{-b}}{e^b - e^{-b}},$$

whereas its limit as $a \to b+$ is

$$\frac{\pi}{2} \cdot \frac{3e^{-b} - e^b}{e^b - e^{-b}},$$

so its value jumps by $-\pi$ as a increases through b.

2.17. The referees of the 1814 memoir were S. F. Lacroix and A. M. Legendre; their report (written by Legendre)[12] was printed together with the memoir when the latter was published in 1827. Legendre raised several questions with Cauchy, whose replies are contained in two supplements attached to the memoir.

Legendre began by challenging Cauchy to evaluate by the methods of his memoir the integral

$$\int_0^{\pi/2} \frac{x\sin 2x \, dx}{a + \cos 2x}.$$

Cauchy successfully evaluated this integral and a number of others that can be dealt with by a similar method.

In his second question, Legendre asked why the condition $a < b$ had to be introduced in establishing the formula

$$\int_0^\infty \frac{\sin ax}{\sin bx} \cdot \frac{dx}{1+x^2} = \frac{\pi}{2} \cdot \frac{e^a - e^{-a}}{e^b - e^{-b}}$$

and its companions in Part II, Section V (Section 2.14 above). Cauchy points out that his argument requires that

$$\frac{\sin a(x + zi)}{\sin b(x + zi)} \cdot \frac{1}{1 + (x + zi)^2}$$

should be small in modulus for large x and z, and that the condition $a < b$

---

[12] Legendre [1814b].

follows from this. He adds some comments on the results obtained in Section VII (Section 2.16 above) by the method of separation of exponentials.

Legendre was not immediately satisfied by Cauchy's explanation; what worried him was that he had himself evaluated[13] the companion integral

$$\int_0^\infty \frac{x \cos ax}{\sin bx} \cdot \frac{dx}{1 + x^2}$$

for the case $a = b$, obtaining the value

$$\frac{\pi}{2} \cdot \frac{2e^{-b}}{e^b - e^{-b}},$$

whereas Cauchy's value for $a < b$ was

$$\frac{\pi}{2} \cdot \frac{e^a + e^{-a}}{e^b - e^{-b}},$$

which would give the different expression

$$\frac{\pi}{2} \cdot \frac{e^b + e^{-b}}{e^b - e^{-b}}$$

if one let a become equal to b. In the second supplement Cauchy derives his own result for $a = b - \alpha$, where α is small and positive, directly from Legendre's expression (which is in fact correct), so that there is no real inconsistency. Similarly, as he shows, the integral

$$\int_0^\infty \frac{\sin ax}{\cos bx} \cdot \frac{dx}{x}$$

has the value $\frac{1}{2}\pi$ when $a = b$, but a value near 0 when a is just less than b.

Cauchy goes on to adduce the integrals

$$\int_0^\infty \frac{\sin ax}{x} \, dx, \qquad \int_0^\infty \frac{x \sin ax}{1 + x^2} \, dx,$$

which both vanish when $a = 0$, but take the values $\frac{1}{2}\pi$ and $\frac{1}{2}\pi e^{-a}$, respectively, when $a > 0$, to indicate that the appearance of these discontinuities has nothing to do with the presence of infinities in the integrand.

Finally, he writes down in detail the values of the integral

$$\int_0^\infty \frac{x \cos ax}{\sin bx} \frac{dx}{1 + x^2}$$

for all $a > 0$, showing that it is discontinuous as a function of a at every odd multiple of b. As often happens, its value at each discontinuity is the mean of its limits on the left and on the right.

---

[13] Legendre [1814a], p. 124.

This further explanation seems to have removed Legendre's remaining doubts.[14]

An unusual phenomenon appears with the integral (another of the family)

$$\int_0^\infty \frac{\cos ax}{\cos bx} \cdot \frac{dx}{1 + x^2}.$$

Its value for $-b < a < b$ is

$$\frac{\pi}{2} \cdot \frac{e^a + e^{-a}}{e^b + e^{-b}},$$

which agrees with the correct value $\frac{1}{2}\pi$ for $a = b$. Nevertheless, the value of the integral as a function of a is discontinuous at $a = b$, its limit as $a \to b+$ being

$$\frac{\pi}{2} \cdot \frac{-e^b + 3e^{-b}}{e^b + e^{-b}};$$

its intervals of continuity are

$$-b \leqslant a \leqslant b, \ b < a < 3b, \ 3b \leqslant a \leqslant 5b, \ 5b < a < 7b, \ldots,$$

so that we have alternately continuity on the left and continuity on the right. In this case the value of the integral at its points of discontinuity fails to coincide with the mean of its left-hand and right-hand limits.

2.18. Let us now try to sum up Cauchy's achievement in the 1814 memoir. As he saw it, he had succeeded in finding a more satisfactory way of obtaining the kind of results for which Euler, Laplace and others had used the method of imaginary substitutions, and at the same time had exhibited some of its limitations. He had created a new set of techniques for the evaluation of definite integrals and was thus in possession of what he was to call a 'calculus', a situation that always aroused his enthusiasm for its exploitation by finding as many and as various applications as he could.

We now list some of the main themes of the memoir, with occasional comments.

   (i) The generalised form of the 'Cauchy–Riemann equations' (Section 2.3). As Cauchy saw it, this was a purely formal result; it was not until 1851 that he came to appreciate the significance of the Cauchy–Riemann equations in distinguishing between analytic and non-analytic functions.
  (ii) A result equivalent to 'Cauchy's theorem' for rectangles and indeed for a class of curvilinear quadrilaterals (Sections 2.3 and 2.4).

[14] Legendre [1814b], p. 609; *Oeuvres de Cauchy* **(1)1**, p. 326.

(iii) His insistence in expressing his results in real terms, and the consequent necessity of his invention of the method of separation of exponentials. He was soon to abandon this restriction, probably by 1817 and certainly by 1819 (Section 3.3 below).

(iv) His discussion of the relation between the definite and indefinite integrals of a function of a single real variable (Section 2.9), where his remarks foreshadow the definitions of a continuous function [1821a], of the principal value of an improper integral [1822b] and of the definite integral as the limit of approximating sums [1823b].

(v) His discussion of double integrals in cases where the two repeated integrals fail to be equal, and his expression of the difference between the two integrals (the 'correction term') as a 'singular integral' (equivalent, in modern terms, to a special kind of repeated limit) (Section 2.11).

(vi) His deeper analysis of the correction term in particular cases, leading to a result equivalent to the usual expression for the residue of a function at a simple pole (Section 2.12). This equivalence is somewhat obscured, since the expression he uses depends on his restriction to functions that are real for real values of the variable. The fact that the result does not depend on the substitution $y = M(x, z) + \mathrm{i}N(x, z)$ introduced in Part I (Section 2.3 above) takes him by surprise.

(vii) His recognition that if a function has several singularities in the domain under consideration, their contributions to the correction terms have to be added together (Section 2.12); in consequence, he is able to state results equivalent to the residue theorem for a special class of contours, in cases where only simple poles are present.

(viii) His numerous evaluations of definite integrals throughout the memoir, especially in Part II. Most of these are infinite integrals, and we have remarked (Sections 2.5 and 2.13) on the inadequacy of his conditions at infinity; he first admitted (tacitly) that these conditions were unsatisfactory in 1823 (Section 3.9 below) but did not replace them by stronger ones till 1825 (Section 4.14 below).

(ix) His examples of infinite integrals that are discontinuous functions of parameters involved in the integrand (Sections 2.16 and 2.17); this discovery probably contributed to Cauchy's attack on the principle of the generality of analysis in the preface to his *Analyse algébrique* [1821a].

(x) His halving of the contributions to the correction term from singularities on the boundary of the domain under consideration

(Section 2.12); this worked very well for simple poles, even though the integrals along the relevant portions of the contour had to be interpreted as principal values, but was to create complications later when he came to consider multiple poles (Sections 4.6 and 5.4 below).

To sum up, although many of the results in the 1814 memoir can be seen to be equivalent to special cases of some of the central theorems of complex function theory, Cauchy himself was still far from appreciating their full significance. It was not until 1825 (Chapter 4 below) that he began to consider integration along complex paths and to introduce some geometrical language into his exposition, and he did not express any of his results in terms of integration round closed contours (circles only at this stage) until 1827 (Section 5.10 below). Nevertheless, his results enabled him to evaluate a great variety of definite integrals, some already known and some quite novel, by methods almost as powerful as, though often clumsier and less perspicuous than, those that one would use today.

3

Miscellaneous contributions (1815–1825)

3.1. Between the completion of his 1814 memoir and his next major contribution to complex function theory, the famous memoir [1825b] on definite integrals between imaginary limits, Cauchy made a number of minor improvements to the theory; in addition, because the publication of the 1814 memoir was held up until 1827, he restated his main results several times, in particular in the papers [1822b] and [1823a] and in his book [1823b] on the infinitesimal calculus.

To get the chronology right, it has to be noted that [1823a] falls into two parts, arising, respectively, from memoirs presented to the Academy of Sciences on 16 September 1822 and 26 May 1823; although the memoir was published in July 1823 and the *Calcul infinitésimal* [1823b] in August, the second part of [1823a] contains references to [1823b]. It is this second part[1] with which we shall mainly be concerned.

We shall also have to take into account some of the supplementary notes to his prize memoir [1815] on water waves, which appear to date from about 1824, and the footnotes that he prepared about 1825 for insertion in the published version of the 1814 memoir. We shall present some evidence about the contents of his 1817 lectures at the Collège de France and an unpublished memoir summarised in [1819]; some relevant material in other papers of this period and in the *Analyse algébrique* will also be mentioned.

Since there are numerous repetitions of material in these publications, it will be most convenient to examine the developments of this period topic by topic rather than discussing the publications in chronological order; we shall give an outline chronology in Section 3.13.

---

[1] The second part is headed 'Observations générales et additions', and concludes with a 'Postscriptum'; it covers pp. 571–92 of [1823a], and pp. 333–57 in the *Oeuvres complètes*, **2(1)**

3.2. We begin by looking at the footnotes added about 1825 to the 1814 memoir. These are mainly concerned with the simplifications of its arguments that came about with the admission of complex integrands. We recall (Section 2.3 above) that in the original version of the 1814 memoir Cauchy used only real integrands; in the footnotes he abandons this restriction altogether. The fundamental equations of Part I, Section I were, in our modification of Cauchy's notation.

$$\int_{x_0}^{X} S(x, Z)\,dx - \int_{x_0}^{X} S(x, z_0)\,dx = \int_{z_0}^{Z} U(X, z)\,dz - \int_{z_0}^{Z} U(x_0, z)\,dz,$$

$$\int_{x_0}^{X} T(x, Z)\,dx - \int_{x_0}^{X} T(x, z_0)\,dx = \int_{z_0}^{Z} V(X, z)\,dz - \int_{z_0}^{Z} V(x_0, z)\,dz$$

(equations (2.3) of Section 2.3 above). As we indicated in Section 2.4, it is natural to combine these into the single equation

$$\int_{x_0}^{X} [S(x, Z) + iT(x, Z)]\,dx - \int_{x_0}^{X} [S(x, z_0) + iT(x, z_0)]\,dx$$

$$= \int_{z_0}^{Z} [U(X, z) + iV(X, z)]\,dx - \int_{z_0}^{Z} [U(x_0, z) + iV(x_0, z)]\,dz, \qquad (3.1)$$

and this is just what Cauchy does in a footnote on the appropriate page.[2] As we saw in Section 2.4, equation (3.1) is essentially equivalent to 'Cauchy's theorem' for a class of curvilinear quadrilaterals.

Cauchy goes on to remark that throughout the memoir every pair of real equations can be replaced by a single imaginary equation. In particular, as we saw in Section 2.4, in the special case given by $M = x$, $N = z$, (3.1) is equivalent to the single equation

$$\int_{x_0}^{X} f(x + iZ)\,dx - \int_{x_0}^{X} f(x + iz_0)\,dx = i\int_{z_0}^{Z} f(X + iz)\,dz - i\int_{z_0}^{Z} f(x_0 + iz)\,dz,$$

$$(3.2)$$

and essentially this equation appears in a footnote[3] in a slightly different notation, and with $x_0 = z_0 = 0$.

This procedure led to important simplifications in the formulae of Part I and in their proofs. We recall, for instance (Section 2.5 above), that in the original version of the memoir Cauchy arrived at four very complicated formulae by taking $f(y)$ of the form $y^{n-1} F(y)$ and working in terms of the real and imaginary parts of $F(y)$; in a footnote to this passage[4] he remarks

---

2 [1814], p. 621; *Oeuvres* **(1)1**, 338.   3 [1814], p. 623; *Oeuvres* **(1)1**, 340.
4 [1814], pp. 628–9; *Oeuvres* **(1)1**, 346–7.

that the four equations can be replaced by a single one, derived by substituting $(y - bi)^{n-1} f(y)$ for $f(y)$ in equations (3.2) above.

The most important simplification in Part I comes in Section VI (Section 2.7 above), where Cauchy had introduced the device he called 'separation of exponentials'. He shows in a footnote[5] that this can be dispensed with very simply by substituting $q(y) e^{ir(y)}$ for $f(y)$ in equation (3.2) above.

There are important simplifications in Part II as well. In Section IV (Section 2.12 above) P' and P'' are defined by

$$f(M \pm Ni) = P' \pm P''i,$$

and S and T by

$$S = P' \frac{\partial M}{\partial x} - P'' \frac{\partial N}{\partial x},$$

$$T = P' \frac{\partial N}{\partial x} + P'' \frac{\partial M}{\partial x}.$$

The corresponding correction term for S is then

$$A = \int_{b'}^{b''} dz \int_{a'}^{a''} \frac{\partial S}{\partial z} \, dx - \int_{a'}^{a''} dx \int_{b'}^{b''} \frac{\partial S}{\partial z} \, dz;$$

similarly for T. Supposing that $f(y)$ is of the form

$$f(y) = \frac{G(y)}{F(y)},$$

where F and G are real for real y, $G(y)$ is well behaved and $F(y)$ has a (simple) zero at $y = \alpha + \beta i$, he defines λ and μ by the equations

$$\lambda = \frac{1}{2} \left[\frac{G(\alpha - \beta i)}{F'(\alpha - \beta i)} + \frac{G(\alpha + \beta i)}{F'(\alpha + \beta i)} \right],$$

$$\mu = \frac{1}{2i} \left[\frac{G(\alpha - \beta i)}{F'(\alpha - \beta i)} - \frac{G(\alpha + \beta i)}{F'(\alpha + \beta i)} \right].$$

He shows that for a zero lying inside the domain of integration we have $A = 2\mu\pi$; the corresponding correction term for T is $2\lambda\pi$. For zeros on the boundary of the domains, the corresponding contribution has to be halved.

In a footnote[6] at this point he concludes that the correction term for $S + Ti$ will therefore be $2\pi i(\lambda - \mu i)$ (with a factor $\frac{1}{2}$ if the zero lies on the boundary). In an earlier footnote,[7] he had already remarked that

$$\lambda - \mu i = \frac{G(\alpha + \beta i)}{F'(\alpha + \beta i)}, \tag{3.3}$$

---

[5] [1814], pp. 647–9; *Oeuvres* **(1)1**, 365–6.   [6] [1814], pp. 697–8; *Oeuvres* **(1)1**, 412–13.
[7] [1814], p. 694; *Oeuvres* **(1)1**, 409.

and that this could also be written as

$$\lambda - \mu i = \varepsilon f(\alpha + \beta i + \varepsilon),\qquad(3.4)$$

where ε is infinitely small (in other words, he is taking the limit as $\varepsilon \to 0$). Although Cauchy does not say so at this stage, it is clear that the hypothesis that $F(y)$ and $G(y)$ are real for real y is no longer playing any role; in fact, he explicitly gets rid of it in a later footnote.[8] Nevertheless, hypotheses such as $f(x - zi) = \overline{f(x + zi)}$ continued to appear sporadically, e.g. in [1822b] and [1823a], but played no essential role in them.

We note that in equations (3.3) and (3.4) above, Cauchy has come very close to his later concept of the residue of a function at a simple pole.

3.3. It is natural to ask when it was that Cauchy decided to abandon the limitation to real-valued integrands that he imposed in the 1814 memoir.

He first used complex integrands in print in a paper [1821b] on the complex form of the Fourier integral, but we can adduce evidence that he had used them in the present context somewhat earlier.

We return to a footnote[9] quoted in Section 3.2. In this he is considering a function of the form

$$f(y) = [\varphi(y) + i\chi(y)]\frac{G(y)}{F(y)},$$

where $G(y)$, $F(y)$, $\varphi(y)$ and $\chi(y)$ are real for real y. He supposes that $f(x + zi)$ vanishes when $x = \pm\infty$ for all $z \geqslant 0$ and when $z = +\infty$ for all x, that $\varphi(x + zi)$, $\chi(x + zi)$, $G(x + zi)$ and $F(x + zi)$ are 'finite and continuous' for $z \geqslant 0$, and that $F(x + zi)$ has (simple) zeros at certain points $\alpha + \beta i$ with $\beta > 0$. Writing

$$\frac{G(\alpha + \beta i)}{F'(\alpha + \beta i)} = \lambda - \mu i, \quad \varphi(\alpha + \beta i) + i\chi(\alpha + \beta i) = \gamma + \delta i,$$

and using a result from an earlier footnote, he derives the equation

$$\int_{-\infty}^{\infty} \varphi(x)\frac{G(x)}{F(x)}\,\mathrm{d}x = 2\pi S(\gamma\mu - \delta\lambda),$$

where the symbol S denotes summation over the zeros $\alpha + \beta i$ with $\beta > 0$. Later in the same footnote he remarks that he had established this equation in his lectures at the Collège de France in 1817; in particular, he had applied it to cases where $\varphi(y) + i\chi(y)$ is of one of the forms

$$\mathrm{e}^{ryi}, \quad \mathrm{i}^{-1}\mathrm{e}^{ryi}, \quad \log(1 - ryi), \quad \mathrm{i}^{-1}\log(1 - ryi).$$

To have obtained the full strength of this result in 1817, Cauchy must

[8] [1814], pp. 712–17; *Oeuvres* **(1)1**, 426–30.
[9] [1814], pp. 712–17; *Oeuvres* **(1)1**, 426–30.

almost certainly have used complex integrands. Our main reason for saying this is that if he had been using methods similar to those of the 1814 memoir, he would have found it necessary either to assume that each of the functions

$$\varphi(x + zi)\frac{G(x + zi)}{F(x + zi)}, \quad \chi(x + zi)\frac{G(x + zi)}{F(x + zi)}$$

satisfied his conditions at infinity separately, or to employ some device like the separation of exponentials, whereas here he makes only the weaker assumption that $f(x + zi)$ satisfies his conditions.

He refers[10] to the 1817 lectures again in 1822, where he quotes, as a new result obtained there, the formula

$$\int_0^\infty x^{a-1} \sin\left(\tfrac{1}{2}a\pi - bx\right)\frac{dx}{1 + x^2} = \tfrac{1}{2}\pi\,e^{-b}(0 < a < 2, \, b > 0).$$

He proves this result in Lesson 39 of the *Calcul infinitésimal* [1823b], using a complex integrand, and he probably obtained it in the same way in 1817.

Whether or not we are right in conjecturing that Cauchy was using complex integrands in 1817, there is little doubt that he was doing so by 1819; in that year he submitted a memoir that was never published (apart from a short and unilluminating summary [1819], which appeared in 1824). In 1822 he quotes[11] from it a formula involving an expression of the form

$$\int_0^\pi \frac{f(\cos p + i\sin p)}{F(\cos p + i\sin p)}\,dp.$$

Cauchy provided a formal justification for the use of complex integrands in Lesson 22 of the *Calcul infinitésimal* [1823b], where he defines the integral of a complex-valued continuous[12] function by means of the equation

$$\int_{x_0}^X (u + vi)\,dx = \int_{x_0}^X u\,dx + i\int_{x_0}^X v\,dx,$$

where u and v are the real and imaginary parts of the integrand.

3.4. Since his name appears in their title, there is perhaps some interest in determining when the 'Cauchy–Riemann' equations made their first appearance in Cauchy's published work. As we have seen (Section 2.3

---

[10] [1822b], p. 161; *Oeuvres* **(2)2**, 295.   [11] [1822b], pp. 168–9; *Oeuvres* **(2)2**, 293.
[12] He had already defined continuity for complex-valued functions of a real variable in [1821a], p. 250; *Oeuvres* **(2)3**, 212.

above), Cauchy introduced in his 1814 memoir a pair of equations that contain the 'Cauchy–Riemann' equations as a special case; however, this memoir did not appear in print until 1827.

The first explicit appearance of these equations in Cauchy's publications seems to have been in the short note [1818]. Here he is considering differential equations of the form

$$\mathrm{d}y - f(x, y)\,\mathrm{d}x = 0,$$

and he shows that if we can find functions $P(x, y)$ and $Q(x, y)$ such that $f(x, y) = Q/P$ (or $f(x, y) = -P/Q$), and $P - Q\mathrm{i}$ is of the form $\varphi(x + y\mathrm{i})$, then P (or Q) will be an integrating factor of the equation. Tacitly assuming, as usual, that $\varphi'(x + y\mathrm{i})$ is meaningful, he deduces that

$$\frac{\partial P}{\partial y} = \frac{\partial Q}{\partial x}, \quad \frac{\partial P}{\partial x} = \frac{\partial(-Q)}{\partial y},$$

essentially the 'Cauchy–Riemann' equations. His result now follows, since $P\,\mathrm{d}x + Q\,\mathrm{d}y$ and $P\,\mathrm{d}y - Q\,\mathrm{d}x$ will be complete differentials. The remainder of the note is not of interest for our purposes.

Another early appearance of the 'Cauchy–Riemann' equations is of some interest. This is in the first part[13] of his 1823 memoir. He supposes that $f(x)$ is real for real x, and writes

$$f(\mu \pm v\mathrm{i}) = M \pm N\mathrm{i}.$$

He wants to calculate what today would be called the Jacobian

$$L = \frac{\partial(M, N)}{\partial(\mu, v)};$$

to do this he writes down the equations

$$\frac{\partial M}{\partial v} = -\frac{\partial N}{\partial \mu}, \quad \frac{\partial N}{\partial v} = \frac{\partial M}{\partial \mu},$$

and deduces at once that

$$L = \left(\frac{\partial N}{\partial \mu}\right)^2 + \left(\frac{\partial N}{\partial v}\right)^2 = f'(\mu + v\mathrm{i})f'(\mu - v\mathrm{i}).$$

Today we should prefer to write

$$L = |f'(\mu + v\mathrm{i})|^2,$$

noting that L is the factor by which area is multiplied in the conformal mapping $\mu + v\mathrm{i} \mapsto f(\mu + v\mathrm{i})$; there is no sign that any such idea was present in Cauchy's mind at the time. The remainder of the passage need not concern us.

---

[13] [1823a], pp. 540ff; *Oeuvres* **(2)1**, pp. 303ff.

3.5. Cauchy restated the basic results of the 1814 memoir in a number of publications. In an 1822 paper[14] he simply states without proof that if $x_0, x_1, \ldots, x_{m-1}$ are the roots of the equation $1/f(x) = 0$ with real parts between x' and x'' and imaginary parts between y' and y'', then

$$\int_{x'}^{x''} [f(x + y''\mathrm{i}) - f(x + y'\mathrm{i})] \, \mathrm{d}x = \mathrm{i} \int_{y'}^{y''} [f(x'' + y\mathrm{i}) - f(x' + y\mathrm{i})] \, \mathrm{d}y$$

$$- 2\pi\mathrm{i}(f_0 + f_1 + \cdots + f_{m-1}),$$

$$(3.5)$$

where $f_0, f_1, \ldots, f_{m-1}$ are the 'true values' of $kf(x_0 + k)$, $kf(x_1 + k), \ldots,$ $kf(x_{n-1} + k)$ for $k = 0$; he has tacitly assumed that all the roots are simple, and he adds his usual remark that contributions from roots on the boundary are to be halved. This result is essentially contained in one of the footnotes[15] added to the 1814 memoir.

Later in the same paper[16] Cauchy gives a more general result. Here he supposes that

$$x = P(p, r) + \mathrm{i}R(p, r),$$

where p and r are real variables, and writes

$$\chi(p, r) = f(P + R\mathrm{i}) \frac{\partial(P + R\mathrm{i})}{\partial p}, \quad \psi(p, r) = f(P + R\mathrm{i}) \frac{\partial(P + R\mathrm{i})}{\partial r};$$

he then gives his basic equations, again without proof, in the form

$$\int_{p'}^{p''} [\chi(p, r'') - \chi(p, r')] \, \mathrm{d}p = \int_{r'}^{r''} [\psi(p'', r) - \psi(p', r)] \, \mathrm{d}r$$

$$- 2\pi\mathrm{i}[\pm f_0 \pm f_1 \pm \ldots \pm f_{n-1}]. \quad (3.6)$$

Here the f_i are the 'true values' of $kf(x_i + k)$ for $k = 0$, where $x_0,$ $x_1, \ldots, x_{n-1}$ are those roots of $1/f(x) = 0$ for which the corresponding values of p and r lie, respectively, between p' and p'' and between r' and r''; the sign $\pm$ in each term on the right-hand side is that of the function

$$\frac{\partial P}{\partial p} \cdot \frac{\partial R}{\partial r} - \frac{\partial P}{\partial r} \cdot \frac{\partial R}{\partial p}$$

(i.e., the Jacobian $\partial(P, R)/\partial(p, r)$) at the corresponding values of (p, r). We recall that in the 1814 memoir[17] he assumed the Jacobian to be positive. As usual, contributions from roots on the boundary are to be halved.

This more general form appears here so that it can be used as a basis for results expressed in polar coordinates (Section 3.7 below).

---

[14] [1822b], p. 167; *Oeuvres* **(2)2**, 291.   [15] [1814], pp. 707–9; *Oeuvres* **(1)1**, 422–3.
[16] [1822b], p. 168; *Oeuvres* **(2)2**, 292.   [17] [1814], p. 696; *Oeuvres* **(1)1**, 411.

Similar restatements of the basic results appear in the second part of the 1823 paper,[18] and in the *Calcul infinitésimal* (cf. Section 3.6 below).

3.6. In Lessons 33–34 of his *Calcul infinitésimal* [1823b] Cauchy gives a new derivation of the basic equation (3.5) above.

He begins with the non-singular case, remarking that if $\varphi(x, y)$ and $\chi(x, y)$ satisfy

$$\frac{\partial \varphi(x, y)}{\partial y} = \frac{\partial \chi(x, y)}{\partial x} = f(x, y). \tag{3.7}$$

then the equation

$$\int_{x_0}^{X} \int_{y_0}^{Y} f(x, y) \, dy \, dx = \int_{y_0}^{Y} \int_{x_0}^{X} f(x, y) \, dx \, dy$$

implies that

$$\int_{x_0}^{X} [\varphi(x, Y) - \varphi(x, y_0)] \, dx = \int_{y_0}^{Y} [\chi(X, y) - \chi(x_0, y)] \, dy, \tag{3.8}$$

provided that $\varphi(x, y)$ and $\psi(x, y)$ are both continuous between the limits of integration. He goes on to say that equation (3.7) will certainly hold if there is a function $f(u)$ such that

$$\varphi(x, y) \, dx + \chi(x, y) \, dy = f(u) \, du,$$

so that

$$\varphi(x, y) = f(u) \frac{\partial u}{\partial x}, \quad \chi(x, y) = f(u) \frac{\partial u}{\partial y}.$$

Furthermore, we may allow $\varphi(x, y)$ and $\chi(x, y)$ to take complex values; in particular, we may take $u = x + y\mathrm{i}$, so that

$$\varphi(x, y) = f(x + y\mathrm{i}), \quad \chi(x, y) = \mathrm{i} f(x + y\mathrm{i}). \tag{3.9}$$

However, equation (3.8) will fail to hold if $\varphi(x, y)$ or $\chi(x, y)$ becomes infinite anywhere in the range of integration; he assumes for the moment that this occurs only when $x = a$ and $y = b$. If ε is small and positive, we then have

$$\int_{x_0}^{a-\varepsilon} [\varphi(x, Y) - \varphi(x, y_0)] \, dx + \int_{a+\varepsilon}^{X} [\varphi(x, Y) - \varphi(x, y_0)] \, dx$$

$$= \int_{y_0}^{Y} [\chi(X, y) - \chi(a + \varepsilon, y) + \chi(a - \varepsilon, y) - \chi(x_0, y)] \, dy.$$

Letting ε tend to 0, we see that (3.8) has to be replaced by

---

18 [1823a], p. 574; *Oeuvres* **(2)1**, 337.

$$\int_{x_0}^{X} [\varphi(x, Y) - \varphi(x, y_0)] \, dx = \int_{y_0}^{Y} [\chi(X, y) - \chi(x_0, y)] \, dy - \Delta, \qquad (3.10)$$

where the correction term Δ is given by

$$\Delta = \lim_{\varepsilon \to 0} \int_{y_0}^{Y} [\chi(a + \varepsilon, y) - \chi(a - \varepsilon, y)] \, dy.$$

He gives an illustrative example, essentially the same as the one in Part II, Section II of the 1814 memoir (Section 2.11 above).

He now returns to the special case $u = x + yi$, and supposes for the moment that $f(u)$ is an (explicit) algebraic function; if $f(u)$ becomes infinite for $u = a + bi$, then equation (3.10) becomes

$$\int_{x_0}^{X} [f(x + Yi) - f(x + y_0 i)] \, dx = i \int_{y_0}^{Y} [f(X + yi) - f(x_0 + yi)] \, dy - \Delta,$$

where

$$\Delta = i \lim_{\varepsilon \to 0} \int_{y_0}^{Y} [f(a + \varepsilon + yi) - f(a - \varepsilon + yi)] \, dy. \qquad (3.11)$$

We note that instead of using the 'Cauchy–Riemann' equations, Cauchy has effectively replaced them by the single equation

$$\frac{\partial}{\partial y} f(x + yi) = i \frac{\partial}{\partial x} f(x + yi).$$

We now come to the chief innovation in the present treatment. Cauchy writes

$$(u - a - bi) f(u) = F(u),$$

and makes the substitution $y = b + \varepsilon z$, denoting the values of z corresponding to y_0 and Y by z_0 and Z. He assumes that both $F(x + yi)$ and $F'(x + yi)$ are finite and continuous, thereby implying that $a + bi$ is a simple zero of $1/f(u)$. Equation (3.11) above now becomes

$$\Delta = i \lim_{\varepsilon \to 0} \int_{z_0}^{Z} \left\{ \frac{F(a + \varepsilon + (b + \varepsilon z)i)}{1 + zi} - \frac{F(a - \varepsilon + (b + \varepsilon z)i)}{-1 + zi} \right\} dz$$

$$= i \lim_{\varepsilon \to 0} \int_{z_0}^{Z} [\varpi(\varepsilon) + i\psi(\varepsilon)] \, dz, \qquad (3.12)$$

say, where ϖ and ψ are also functions of z. He then writes

$$\alpha = \frac{\varpi(\varepsilon) - \varpi(0)}{\varepsilon}, \qquad \beta = \frac{\psi(\varepsilon) - \psi(0)}{\varepsilon},$$

and remarks that, since

$$\varpi'(\varepsilon) + i\psi'(\varepsilon) = F'(a + \varepsilon + yi) - F'(a - \varepsilon + yi),$$

both $\varpi'(\varepsilon)$ and $\psi'(\varepsilon)$ are small when ε is small; by the mean-value theorem, the same is true of α and β. He then rewrites (3.12) in the form

$$\Delta = i\lim_{\varepsilon \to 0}\int_{z_0}^{Z}(\alpha + i\beta)\varepsilon\,dz + i\lim_{\varepsilon \to 0}\int_{z_0}^{Z}[\varpi(0) + i\psi(0)]\,dz.$$

He claims that the first term on the right-hand side must vanish, so that

$$\Delta = i\lim_{\varepsilon \to 0}\int_{z_0}^{Z}[\varpi(0) + i\psi(0)]\,dz.$$

Since $\lim z_0 = -\infty$ and $\lim Z = +\infty$, it follows that

$$\Delta = i\int_{-\infty}^{\infty}[\varpi(0) + i\psi(0)]\,dz$$

$$= i\int_{-\infty}^{\infty}F(a+bi)\left(\frac{1}{1+zi} - \frac{1}{-1+zi}\right)dz$$

$$= 2iF(a+bi)\int_{-\infty}^{\infty}\frac{dz}{1+z^2}$$

$$= 2\pi i f_0,$$

where $f_0 = F(a+bi) = \lim \varepsilon(a+bi+\varepsilon)$.

This argument is a good deal simpler than the one given in the 1814 memoir (Section 2.12 above). The main reasons for this are that (i) he has confined his attention to the case $u = x + yi$, and has not considered any more general form for u, and (ii) he has dealt directly with complex-valued functions instead of treating their real and imaginary parts separately. His argument is still some way from satisfying modern canons of rigour, however; his conclusion that

$$\lim_{\varepsilon \to 0}\int_{z_0}^{Z}\varepsilon(\alpha + i\beta)\,dz = 0$$

requires more justification than he gives it, in view of the fact that $z_0 \to -\infty$ and $Z \to +\infty$.

He continues with the standard remarks that if the equation $f(u) = \infty$ has m roots in the range of integration, then

$$\Delta = 2\pi i(f_1 + f_2 + \cdots + f_m). \tag{3.13}$$

where f_r is the 'true value' of $(u - u_r)f(u)$ at the root u_r, and that contributions from roots on the boundary have to be halved.

Cauchy then gives his usual application of the result, i.e. that if $f(x + yi)$ vanishes when $x = \pm\infty$ for all y and when $y = +\infty$ for all x, then

$$\int_{-\infty}^{\infty} f(x)\,dx = \Delta, \tag{3.14}$$

where the sum in equation (3.13) is taken over the infinities of $f(x + yi)$ for which $y > 0$.

His conditions at infinity are still inadequate, in spite of the fact that he had given correct criteria for the convergence of infinite integrals in Lesson 25 (cf. Section 3.9 below).

Among his illustrative examples is the formula

$$\int_{-\infty}^{\infty} \frac{f(x)}{1 + x^2}\,dx = \pi f(i), \tag{3.15}$$

applying to functions $f(x + yi)$ with no infinities in $y > 0$. As a special case, he takes

$$f(u) = (-ui)^{\mu-1},$$

where $0 < \mu < 2$, and so derives the equation

$$\int_0^{\infty} \frac{x^{\mu-1}\,dx}{1 + x^2} = \frac{\pi}{2\sin\frac{1}{2}\mu\pi}.$$

We recall that in the 1814 memoir he had obtained an equivalent result by a somewhat dubious argument (Section 2.14 above).

In Lesson 39 of the *Calcul infinitésimal* Cauchy removes the restriction to algebraic functions that he had imposed in Lesson 34, indicating (without further proof) that equation (3.15) above will continue to hold for functions $f(u)$ such as

$$e^{iau}, \quad (-ui)^{\mu-1} \cdot e^{iau}, \quad \frac{(-ui)\,e^{iau}}{\log(1 - iru)},$$

where $0 < \mu < 2$, $a > 0$, $r > 0$, and he gives a number of illustrative examples.

3.7.   We recall (Section 2.6 above) that in Part I, Section IV of the 1814 memoir Cauchy considered the special case defined by

$$M = x\cos z, \quad N = x\sin z,$$

so that, in geometrical terms, he was transforming to polar coodinates. In Part II, however, he gave this case no special attention, although the general results of Section IV (Section 2.12 above) would be applicable to it.

He repairs this omission in [1822b] and [1823a]. The special form of equation (3.6) for this case first appears explicitly in the latter paper,[19] in the form

---

[19]   [1823a], p. 575; *Oeuvres* (2)1, 338.

$$\int_{u'}^{u''} [e^{v''i} f(u\, e^{v''i}) - e^{v'i} f(u\, e^{v'i})]\, du$$

$$= i \int_{v'}^{v''} [u'' f(u''\, e^{vi}) - u' f(u'\, e^{vi})]\, dv - 2\pi i (f_0 + f_1 + \cdots + f_{m-1}).$$

In [1822b] he quotes[20] from the unpublished 1819 memoir the formula

$$\int_0^\pi \frac{f(\cos p + i \sin p)}{F(\cos p + i \sin p)}\, dp$$

$$= i \int_{-1}^1 \frac{f(r)}{rF(r)}\, dr + \pi \left\{ \frac{f(0)}{F(0)} + \frac{f(a)}{aF'(a)} + \frac{f(a')}{a'F'(a')} + \cdots \right\}$$

$$+ 2\pi \left\{ \frac{f(\alpha + \beta i)}{(\alpha + \beta i)F'(\alpha + \beta i)} + \frac{f(\alpha' + \beta' i)}{(\alpha' + \beta' i)F'(\alpha' + \beta' i)} + \cdots \right\}, \qquad (3.16)$$

where $a, a', \ldots$ are the real roots of $F(x) = 0$ with absolute value less than 1, and $\alpha + \beta i$, $\alpha' + \beta' i$, $\ldots$ are the non-real roots with absolute value less than 1 and positive imaginary part. As he says, this result can be derived from equation (3.6) by replacing $f(x)$ by $f(x)/xF(x)$ and taking

$$P + Ri = r(\cos p + i \sin p),$$

with $p' = 0$, $p'' = \pi$, $r' = 0$, $r'' = 1$. He tacitly assumes, as usual, that the zeros of $F(x)$ are all simple, and also that $F(0) \neq 0$.

In [1823a] Cauchy replaces (3.16) by the simpler equation

$$\int_{-\pi}^\pi \frac{f(e^{pi})}{F(e^{pi})}\, dp = 2\pi \left\{ \frac{f(0)}{F(0)} + \frac{f(x_0)}{x_0 F'(x_0)} + \frac{f(x_1)}{x_1 F'(x_1)} + \cdots \right\}, \qquad (3.17)$$

where $x_0, x_1, \ldots$ are the roots of $F(x) = 0$ with absolute value less than 1. The distinction between real and non-real roots has disappeared, but the assumption that $F(0) \neq 0$ remains. The special treatment of the centre of the circle survived for a long time (Sections 5.8 and 5.10 below).

We note that (3.16) is equivalent to a case of the residue theorem for the semicircular domain bounded by the upper half of the unit circle and the interval $[-1, 1]$ of the real axis, the integral on the right-hand side of (3.16) being interpreted as a principal value if necessary (cf. Section 3.8 below), and equation (3.17) is a case of the residue theorem for the unit circle.

In the 1822 paper, Cauchy quotes from the 1819 memoir, as a special case of (3.16) above, the formula[21]

---

[20] [1822b], p. 169; *Oeuvres* (2)2, 293; note that, in the *Oeuvres* equation (25), $F'(0)$ should read $F(0)$; this is correct in the original paper, but there $F(a)$ should be $F'(a)$.
[21] [1822b], p. 169; *Oeuvres* (2)2, 293.

$$\int_0^\pi \frac{e^{pi} f(e^{pi})}{e^{pi} - a} \, dp = \pi f(a) + \int_{-1}^1 \frac{f(r)}{r - a} \, dr, \qquad (3.18)$$

where $0 < a < 1$. From this, he says, without giving any proofs, one can easily deduce the equations

$$\frac{1}{2} \int_{-\pi}^\pi e^{-npi} f(b + e^{pi}) \, dp = \frac{\pi}{n!} f^{(n)}(b), \qquad (3.19)$$

$$\frac{1}{2} \int_{-\pi}^\pi e^{-npi + he^{pi}} \, dp = \frac{\pi h^n}{n!}, \qquad (3.20)$$

$$\Delta^m h^n = \frac{n!}{2\pi} \int_{-\pi}^\pi (e^{e^{pi}} - 1)^m e^{-npi + he^{pi}} \, dp, \qquad (3.21)$$

where m and n are positive integers, and b and h are arbitrary constants (presumably real, although he does not say so).

We note that (3.19) is essentially the familiar result expressing the derivatives of an analytic function at the centre of a circle in terms of its values on the circumference; it was not until about 1831 (cf. Section 6.9 below) that Cauchy perceived how powerful a tool this equation could be.

Cauchy goes on to list a further chain of results, which he obtains by taking real parts in (3.18), tacitly assuming that $f(u)$ takes conjugate values for conjugate values of u. The first of these results is

$$\frac{1}{2} \int_0^\pi \left\{ \frac{f(e^{pi})}{1 - ae^{-pi}} + \frac{f(e^{-pi})}{1 - ae^{pi}} \right\} dp = \pi f(a) \qquad (0 < a < 1), \qquad (3.22)$$

which is equivalent to a special case of 'Cauchy's formula' for the unit circle. Returning to (3.16), and taking $F(r) = 1 - ar$, still with $0 < a < 1$, he derives

$$\frac{1}{2} \int_0^\pi \left\{ \frac{f(e^{pi})}{1 - ae^{pi}} + \frac{f(e^{-pi})}{1 - ae^{-pi}} \right\} = \pi f(0). \qquad (3.23)$$

He also gives the corresponding results for $a > 1$; these are

$$\frac{1}{2} \int_0^\pi \left\{ \frac{f(e^{pi})}{1 - ae^{-pi}} + \frac{f(e^{-pi})}{1 - ae^{pi}} \right\} dp = 0, \qquad (3.24)$$

$$\frac{1}{2} \int_0^\pi \left\{ \frac{f(e^{pi})}{1 - ae^{pi}} + \frac{f(e^{-pi})}{1 - ae^{-pi}} \right\} dp = \pi \left[f(0) - f\left(\frac{1}{a}\right) \right]. \qquad (3.25)$$

He remarks that by adding and subtracting (3.22) and (3.23) one can obtain two equations recently announced by Poisson.[22] These are

$$\frac{1}{\pi} \int_0^\pi \frac{[F(a + e^{-xi}) + F(a + e^{xi})](1 - p\cos x)}{1 - 2p\cos x + p^2} \, dx = F(a + p) + F(a),$$

---

[22] Poisson [1822], p. 138.

$$\frac{\mathrm{i}p}{\pi} \int_0^\pi \frac{[F(a + \mathrm{e}^{-xi}) - F(a + \mathrm{e}^{xi})] \sin x}{1 - 2p \cos x + p^2} \, \mathrm{d}x = F(a + p) - F(a),$$

where $0 < p < 1$.

Cauchy returns to some of these results in his 1823 memoir,[23] but expresses them more succinctly, (3.23) and (3.25) being replaced by

$$\int_{-\pi}^\pi \frac{f(\mathrm{e}^{vi})}{1 - a \, \mathrm{e}^{vi}} \, \mathrm{d}v = 2\pi f(0) \qquad (a^2 < 1),$$

$$\int_{-\pi}^\pi \frac{f(\mathrm{e}^{vi})}{1 - a \, \mathrm{e}^{vi}} \, \mathrm{d}v = 2\pi \left[f(0) - f\left(\frac{1}{a}\right) \right] \qquad (a^2 > 1).$$

To a modern eye it looks as if Cauchy is integrating round a complete circle; however, it does not appear that he began to think in these terms until 1827 (cf. Section 5.10 below). At this stage he was not using any geometrical language.

There is an interesting postscript[24] to the 1822 paper. Assuming[25] that the Maclaurin series

$$f(0) + \frac{x}{1} f'(0) + \frac{x^2}{2!} f''(0) + \cdots$$

is convergent to $f(x)$ for $|x| < 1$ (actually he needs this for $|x| \leq 1$), he shows that equations (3.19) to (3.25) inclusive can be obtained by applying Parseval's theorem, which he quotes in the following form: if

$$\varphi(x) = a_0 + a_1 x + a_2 x^2 + \cdots,$$
$$\chi(x) = b_0 + b_1 x + b_2 x^2 + \cdots,$$
$$\psi(x) = a_0 b_0 + a_1 b_1 x + a_2 b_2 x^2 + \cdots,$$

then

$$\psi(xy) = \frac{1}{2\pi} \int_0^\pi [\varphi(x \, \mathrm{e}^{pi})\chi(y \, \mathrm{e}^{-pi}) + \varphi(x \, \mathrm{e}^{-pi})\chi(y \, \mathrm{e}^{pi})] \, \mathrm{d}p$$

$$= \frac{1}{2\pi} \int_{-\pi}^\pi \varphi(x \, \mathrm{e}^{pi})\chi(y \, \mathrm{e}^{-pi}) \, \mathrm{d}p. \tag{3.26}$$

This is an immediate consequence of the theorem as originally stated by Parseval [1805], which was[26]

$$\pi\psi(1) = \frac{1}{2} \int_0^\pi [\varphi(\mathrm{e}^{pi})\chi(\mathrm{e}^{-pi}) + \varphi(\mathrm{e}^{-pi})\chi(\mathrm{e}^{pi})] \, \mathrm{d}p.$$

---

[23] [1823a], p. 579; *Oeuvres* **(2)1**, 343.   [24] [1822b], pp. 173–4; *Oeuvres* **(2)2**, 297–9.
[25] There are several obvious misprints in his statement of this hypothesis in [1822b]; one of them persists in *Oeuvres* (p. 297).
[26] In this equation $\psi(1)$ is misprinted in the *Oeuvres* as $\chi(1)$; it is correct in the original paper.

To establish (3.19) to (3.25) he simply indicates for each equation[27] the values to be given to $\varphi(x)$ and $\chi(x)$ in (3.26).

Frullani [1818] (published 1820) had also used Parseval's theorem, to show that if

$$f(x) = a_0 + a_1 x + a_2 x^2 + \cdots$$

is real for real x, then

$$f(0) = a_0 = \frac{1}{2\pi} \int_0^\pi [f(e^{\varphi i}) + f(e^{-\varphi i})] \, d\varphi,$$

$$\frac{f^{(m)}(0)}{n!} = a_n = \frac{1}{\pi} \int_0^\pi [f(e^{\varphi i}) + f(e^{-\varphi i})] \cos n\varphi \, d\varphi.$$

Cauchy does not mention Frullani's paper here, but does cite it in a later paper [1826g].

3.8. In the present section we discuss the formal introduction by Cauchy of principal-value integrals; although this notion is not essential to the exposition of complex analysis, it did play a considerable role in Cauchy's treatment of the subject.

We recall that in the 1814 memoir he discussed definite integrals in cases where the integrand has an infinity in the range of integration (Section 2.9 above), that he gave an argument leading to an interpretation equivalent to taking the principal value and that many of the special infinite integrals he evaluated (Sections 2.14 and 2.16 above) had to be interpreted in this way. We also recall that in his evaluation of correction terms (Section 2.13 above) he indicated that the contribution from a singularity on the boundary of the domain is to be halved; the integral along this part of the boundary is then to be treated as a principal value.

He gives a detailed exposition[28] of these ideas in his 1822 paper. In the 1814 memoir he had defined a *singular integral*[29] as an integral taken between limits that are infinitely close to one another. In the 1822 paper he extends this definition to cover integrals, with respect to one or more variables, between limits that are infinitely close to given values, which may themselves be infinite. Thus, supposing that $f(x)$ has an infinity at $x = x_0$, that f_0 is the 'true value' of $kf(x_0 + k)$ for $k = 0$, i.e. in modern notation,

$$f_0 = \lim_{k \to 0} kf(x_0 + k),$$

---

[27] The values to be taken for equation (3.24) are erroneous, both in the original and in the *Oeuvres*; for this case one must take $\varphi(x) = -xf(x)/(a - x)$, $\chi(x) = 1$.

[28] [1822b], pp. 162–6; *Oeuvres* **(2)2**, 284–90.   [29] [1814], p. 678; *Oeuvres* **(1)1**, 394.

and that α' and α'' are positive constants, he says that the singular integral

$$\int_{x_0+k\alpha'}^{x_0+k\alpha''} f(x)\,dx$$

will become equal to $f_0 \log(\alpha'/\alpha'')$ when k is infinitely small. He also considers similar expressions involving double and triple integrals, but these need not concern us.

He also considers singular integrals such as

$$\int_{1/k\alpha'}^{1/k\alpha''} f(x)\,dx,$$

which, he says, becomes equal to $f_0 \log(\alpha''/\alpha')$, where f_0 is the 'true value' of $f(1/k)/k$ for $k = 0$, or, equivalently, the 'true value' of $xf(x)$ for $x = \infty$.

More generally, if $f(x)$ is infinite or indeterminate at the values x_0, $x_1, \ldots, x_{n-1}$ in the interval (x', x''), then the integral

$$\int_{x'}^{x''} f(x)\,dx$$

is practically ('sensiblement') equivalent to an expression of the form

$$A = \int_{x'}^{x_0-k\alpha'} f(x)\,dx + \int_{x_0+k\alpha''}^{x_1-k\beta'} f(x)\,dx + \cdots + \int_{x_{m-1}+k\varepsilon''}^{x''} f(x)\,dx,$$

where k is very small. In particular, if one takes

$$\alpha' = \alpha'' = \beta' = \beta'' = \cdots = \varepsilon' = \varepsilon'' = 1,$$

with k infinitely small, one obtains what he proposes to call the *principal value* of the integral; this is usually unique, although it may be infinite. To obtain the 'general value' of the integral one has to add on the values of the singular integrals

$$\int_{x_0-k}^{x_0-k\alpha'} f(x)\,dx, \quad \int_{x_0+k\alpha''}^{x_0+k} f(x)\,dx, \ldots, \quad \int_{x_{n-1}+k\varepsilon''}^{x_{n-1}+k} f(x)\,dx;$$

the result will thus usually be of the form

$$A = B + f_0 \log \frac{\alpha'}{\alpha''} + f_1 \log \frac{\beta'}{\beta''} + \cdots + f_{n-1} \log \frac{\varepsilon'}{\varepsilon''},$$

where B is the principal value, and $f_0, f_1, \ldots, f_{n-1}$ are the 'true values' of

$$(x - x_0)f(x), (x - x_1)f(x), \ldots, (x - x_{n-1})f(x)$$

for $x = x_0, x_1, \ldots, x_{n-1}$.

No serious use of this notion of the 'general value' of an improper integral seems to have been made by Cauchy or anyone else; on the other hand, the concept of 'principal value' has survived and has proved to be useful in numerous parts of analysis.

Later in the 1822 paper Cauchy makes some remarks[30] about definite integrals in general. He insists that a definite integral between real limits should always be thought of as the sum of the values of the differential corresponding to the values of the real variable between the limits of integration. This point of view is to be preferred, he says, since it can still be used even when no primitive function is known; also, it always gives real values for the integrals of real functions, and it allows one to separate every complex equation into two real equations. None of this holds if the definite integral is considered as necessarily equivalent to the difference between two values of a (possibly discontinuous) primitive function, or if one passes from one limit to the other through complex values.

These remarks were provoked by a paper published by Poisson [1820], in which he obtained results such as

$$\int_0^\infty \frac{\cos ax \, dx}{x^2 - b^2} = -\frac{\pi}{2b} \sin ab - \frac{1}{2b} \cos ab \log(-1),$$

giving an imaginary value for a real integral, and tried to justify them by imaginary changes of variable; we shall have more to say about this paper in Section 4.2 below. Cauchy points out that the principal value of this integral is

$$\frac{-\pi \sin ab}{2b}$$

and that its 'general value' is of the form

$$\frac{\pi}{2b} (\cos ab \log m - \sin ab).$$

This investigation of Cauchy's may have played some part in clarifying his ideas about the nature of the definite integral; his formal definition of the integral as a limit of approximating sums was to come in Lesson 21 of the *Calcul infinitésimal* [1823b]. His use of the phrase 'between real limits' possibly foreshadows his famous memoir [1825b] on definite integrals between imaginary limits (Chapter 4 below).

It may also be mentioned here that it was in [1822b] that Cauchy adopted the now standard notation

$$\int_a^b f(x) \, dx$$

[30] [1822b], p. 171, *Oeuvres* (2)2, 295–6.

for definite integrals, with the limits of integration attached to the integral
sign, a notation that had recently been introduced by Fourier.

Cauchy returns to the topic of singular integrals, principal values and
general values in the *Calcul infinitésimal* [1823b]; on this occasion,
however, he deals with general values and principal values first, in Lesson
24, deferring singular integrals to Lesson 25. Some of the material on these
topics reappears in the later part of [1823a]. Here he makes one new point,
remarking that many indeterminate integrals may be regarded as limits of
integrals containing a parameter that ultimately becomes 0. Among his
illustrative examples he gives

$$\int_0^m \frac{dx}{x-1} = \lim_{k \to 0} \int_0^m \frac{(x-1)\,dx}{k^2+(x-1)^2} = \lim_{k \to 0} \frac{1}{2} \log \frac{k^2+(m-1)^2}{k^2+1} = \log(m-1).$$

He also makes some remarks about the conversion of indeterminate
integrals to ordinary ones, extending his discussion in Part II, Section V of
the 1814 memoir of Legendre's device (Section 2.14 above).

3.9. We recall that Cauchy's handling of infinite integrals in the 1814
memoir is unsatisfactory (Sections 2.5, 2.7 and 2.13 above), since the conditions he imposes are inadequate to ensure the validity of his arguments or
even the convergence of the integrals. He continues to use similar conditions throughout the period dealt with in the present chapter on a number
of occasions; however, inconsistencies begin to manifest themselves.

The question usually arises when he is obtaining a formula such as

$$\int_{-\infty}^{\infty} f(x)\,dx = 2\pi i (f_1 + f_2 + \cdots + f_n),$$

where, if $x_1, x_2, \ldots, x_n$ are the infinities of $f(x)$ with positive imaginary
part, f_k $(k = 1, \ldots, n)$ is the 'true value' of $(x - x_k)f(x)$ at $x = x_k$; his
standard conditions are that $f(x + yi)$ should vanish for all y when
$x = \pm\infty$ and for all x when $y = +\infty$. These conditions appear in Lesson
34 of the *Calcul infinitésimal* [1823b] and in the 1823 paper;[31] also in the
Notes[32] added in 1824 to his prize essay on waves, and in a footnote[33]
(dating from about 1825) to the 1814 memoir.

On the other hand, in Lesson 24 of the *Calcul infinitésimal* [1823b],
Cauchy defines

$$\int_{x_0}^{\infty} f(x)\,dx$$

[31] [1823a], p. 575; *Oeuvres* **(2)1**, 338.   [32] [1815], p. 291; *Oeuvres* **(1)1**, 283.
[33] [1814], p. 713; *Oeuvres* **(1)1**, 427.

as the limit when X becomes infinite of

$$\int_{x_0}^{X} f(x)\,dx;$$

and he defines the principal value of

$$\int_{-\infty}^{\infty} f(x)\,dx$$

as the limit when X becomes infinite of

$$\int_{-X}^{X} f(x)\,dx.$$

At the end of Lesson 25, he says that if $f(x)/F(x)$ is a rational function, then

$$\int_{-\infty}^{\infty} \frac{f(x)}{F(x)}\,dx$$

will be finite and determinate if and only if (i) the equation $F(x) = 0$ has no real roots, and (ii) the degree of $F(x)$ exceeds that of $f(x)$ by at least 2; this, of course, is correct. He returns to this topic in Lesson 32, where he gives the following result. Let $f(x)/F(x)$ be a rational function (real for real x) such that $F(x) \neq 0$ for real x; if x_1, x_2 ... are the zeros (tacitly supposed simple) of $F(x)$ with positive imaginary part, and $A_1 - B_1\mathrm{i}$, $A_2 - B_2\mathrm{i}$, ... are the values of $f(x)/F'(x)$ at these zeros,[34] then

$$\lim \int_{-1/\varepsilon\mu}^{1/\varepsilon\nu} \frac{f(x)}{F(x)}\,dx = 2\log\left(\frac{\mu}{\nu}\right)(A_1 + A_2 + \cdots) + 2\pi(B_1 + B_2 + \cdots),$$

the limit being taken as ε becomes 0. In particular, if the degree of $F(x)$ exceeds that of $f(x)$ by at least 2, then $A_1 + A_2 + \cdots = 0$, and we have

$$\int_{-\infty}^{\infty} \frac{f(x)}{F(x)}\,dx = 2\pi(B_1 + B_2 + \cdots). \tag{3.27}$$

If the degree of $F(x)$ exceeds that of $f(x)$ by unity, then (3.27) still holds, provided that the integrand is interpreted as a principal value; one can also allow $F(x)$ to have real zeros if a principal-value interpretation is used.

An inconsistency appears in Cauchy's 1823 paper, where he uses his usual inadequate standard conditions at infinity. He starts from the equation

$$\frac{f(x_0)}{F'(x_0)} + \frac{f(x_1)}{F'(x_1)} + \cdots + \frac{f(x_{m-1})}{F'(x_{m-1})} = \frac{1}{2\pi}\int_{-\infty}^{\infty} \left\{\frac{f(u''+iv)}{F(u''+iv)} - \frac{f(u'+iv)}{F(u'+iv)}\right\} dv,$$

---

[34] $f(x)/F'(x)$ is misprinted as $f(x)/F(x)$, both in the original, p. 126, and in *Oeuvres* (2)4, 190.

where $f(u + iv)/F(u + iv)$ vanishes when $v = \pm\infty$ for all values of u, and $x_0, x_1, \ldots, x_{m-1}$ are the zeros of $F(x)$ with real parts between u' and u''. He examines the particular case where

$$\frac{f(x)}{F(x)} = \frac{a_0 + a_1x + \cdots + a_px^p}{A_0 + A_1x + \cdots + A_mx^m}.$$

If $p < m - 1$, he takes $u' = -\infty$, $u'' = +\infty$, and obtains

$$\frac{f(x_0)}{F'(x_0)} + \frac{f(x_1)}{F'(x_1)} + \cdots + \frac{f(x_{m-1})}{F'(x_{m-1})} = 0, \tag{3.28}$$

the sum now being over all the zeros of $F(x)$. If, however, $p = m - 1$, he takes $u' = -U$, $u'' = U$, where U is large and positive, so that the sum is equal to

$$\frac{1}{2\pi} \int_{-\infty}^{\infty} \left\{ \frac{f(U + iv)}{F(U + iv)} - \frac{f(-U + iv)}{F(-U + iv)} \right\} dv. \tag{3.29}$$

He then argues that the expression (3.29) is approximately equal to

$$\frac{1}{2\pi} \int_{-\infty}^{\infty} \frac{a_{m-1}}{A_m} \left(\frac{1}{U + iv} - \frac{1}{-U + iv} \right) dv = \frac{a_{m-1}}{\pi A_m} \int_{-\infty}^{\infty} \frac{U\, dv}{U^2 + v^2}$$

$$= \frac{a_{m-1}}{A_m},$$

concluding that in this case we have

$$\frac{f(x_0)}{F'(x_0)} + \frac{f(x_1)}{F'(x_1)} + \cdots + \frac{f(x_{m-1})}{F'(x_{m-1})} = \frac{a_{m-1}}{A_m}. \tag{3.30}$$

The results (3.28) and (3.30), as he remarks, agree with those he obtained, by a purely algebraic method, in an earlier paper.[35]

We note that his standard conditions would have given him an erroneous result; in the expression (3.29) the integrand vanishes when U becomes infinite, so that, by his usual argument, the integral should have the limit 0. In fact, equation (3.30) is correct; he has thus recognised, although not explicitly, that his usual argument does not always give the right answer. Cauchy's attention was almost certainly drawn to the discrepancy when he compared his results with the purely algebraic conclusions of [1813]. He discussed this inconsistency again in the 1825 memoir (Section 4.11 below) and still further in [1825c] (Section 4.14 below).

In modern language, what Cauchy has discovered here is the residue of $f(x)/F(x)$ at infinity. The discussions in Lesson 25 and 32 of the *Calcul infinitésimal* and mentioned earlier in this section point in the same direction, but rather less clearly.

---

[35] [1813], pp. 497–500; *Oeuvres* (2)**1**, 207–10.

3.10. Cauchy remarked in Part II, Section IV of the 1814 memoir (Section 2.12 above) that his formulae for the correction term became meaningless in cases where multiple roots were involved; he promised to examine such cases later, but he did not do so either in the 1814 memoir or in the footnotes he added to it before its publication in 1827, nor did he mention the matter in the *Calcul infinitésimal* [1823b].

However, in a footnote[36] to the 1823 paper he states, but without proof, that if the equation $f(x) = \pm\infty$ has a multiple root of order p at $x = x_0$, then the value of f_0, where $2\pi i f_0$ is the contribution of this root to the correction term, is no longer the 'true value' of $kf(x_0 + k)$ when $k = 0$, but now has to be the 'true value' of

$$\frac{1}{(p-1)!} \frac{\mathrm{d}^{p-1}[k^p f(x_0 + k)]}{\mathrm{d}k^{p-1}}$$

when $k = 0$. In effect, we have here an expression for the residue of a function at a multiple pole.

His methods for establishing this result will be discussed in Chapter 4 below.

3.11. During the period with which we are concerned Cauchy paid considerable attention to the solution of algebraic (and, later, transcendental) equations. He studied questions about the existence of solutions, about the localisation of roots, and about exact formulae for them.

He gave several proofs of what has become known as the fundamental theorem of algebra, to the effect that every non-constant polynomial has at least one zero. His first proof [1817a] has some similarity to Gauss's first proof [Gauss 1799]. Cauchy considers separately the zeros of the real and imaginary parts of the polynomial $f(z)$ when $|z| = R$, where R is initially a large constant, and then discusses the behaviour of these zeros as R is gradually diminished.

His second, more analytical, proof appeared in [1817b]. He proves first that $|f(z)|^2$ becomes infinite with $|z|$, whence every zero z must satisfy $|z| \leqslant R$ for some R; taking for granted that $|f(z)|^2$ must assume its lower bound somewhere in this region, he shows that its minimum value has to be 0.

In both these proofs he considers only polynomials with real coefficients, and the same is true of his republication of the second proof in [1820]. He remarks at one point that the method would apply to any function $f(z)$ for

---

[36] [1823a], p. 574; *Oeuvres* (2)**1**, 337.

which $|f(z)|^2$ becomes infinite with $|z|$; this remark may have led him to give the result in his *Analyse algébrique*[37] for polynomials with complex coefficients, with essentially the same proof.

It should be mentioned that Argand had given a proof of the theorem on lines similar to Cauchy's second proof [Argand 1815]. Several points about this proof deserve notice. It is semi-geometrical, being based on Argand's geometrical representation of the complex numbers; he fails to perceive the necessity of proving first that $|f(z)|$ becomes infinite with $|z|$; his proof that the minimum attained by $|f(z)|$ is 0 is essentially a geometrical version of Cauchy's; and his proof is given for polynomials with complex coefficients.

The first proof for the case of complex coefficients is sometimes attributed[38] to Gauss [1849]. It now appears that both Argand and Cauchy deserve some credit for the proof of this case. How much Cauchy knew of Argand's contributions, and when he learned of them, is a question still in debate.

In an early paper [1813] Cauchy gave some results on the number of roots of an algebraic equation in a given interval; he was to return to problems about the localisation of roots, both in real and in complex contexts, at a later stage (see especially Chapter 6 below).

In [1819] Cauchy accepted the validity of Ruffini's (almost complete) proof that the general equation of the fifth degree cannot be solved in terms of radicals. Thus there was to his mind little point in pursuing this line of attack, and he began to seek other ways of finding explicit formulae for the solution of equations. In the published abstract of [1819] he announced that he had found a method by which each root of an algebraic equation could be expressed as a definite integral whose integrand is a rational function. He gives an account[39] of his results in this direction in his 1823 paper, where he considers the more general problem of finding explicit formulae, not only for the roots themselves, but for various expressions involving the roots. If he wishes to study the roots of the equation $F(x) = 0$, he forms a function of the form

$$f(x) = \frac{\varphi(x)F'(x) - \varpi(x)F(x)}{F(x)},$$

where $\varphi(x)$ and $\varpi(x)$ are well-behaved functions.

From his general theory he derives the equation

[37] [1821a], pp. 329ff; *Oeuvres* **(2)3**, 274ff.   [38] See, for example, Kline [1972], p. 599.
[39] [1823a], pp. 579–82; *Oeuvres* **(2)1**, 343–6.

$$\varphi(x_0) + \varphi(x_1) + \cdots + \varphi(x_{m-1}) = \frac{1}{2\pi} \int_{v'}^{v''} [f(u'' + vi) - f(u' + vi)] \, dv$$

$$+ \frac{i}{2\pi} \int_{u'}^{u''} [f(u + v''i) - f(u + v'i)] \, du,$$

$$(3.31)$$

where $x_0, x_1, \ldots, x_{m-1}$ are the zeros (tacitly supposed simple) of $F(x)$ with real parts between u' and u'' and imaginary parts between v' and v''. He gives a similar equation for the case of polar coordinates; an important special case of this is

$$\varphi(x_0) + \varphi(x_1) + \cdots + \varphi(x_{m-1}) = \frac{1}{2\pi} \int_{-\pi}^{+\pi} r \, e^{vi} f(r \, e^{vi}) \, dv, \qquad (3.32)$$

where $x_0, x_1, \ldots, x_{m-1}$ are the zeros of $F(x)$ with modulus less than r. Cauchy was to make very effective use of formulae like (3.32) in his later investigations of the convergence of the Lagrange and other related expansions for implicit functions (Chapter 6 below).

Cauchy remarks that if the limits of integration in (3.31) are chosen appropriately there will be only one relevant zero of $F(x)$, say x_0; we can thus obtain an explicit integral expression for $\varphi(x_0)$, and in particular for x_0 itself. He mentions that these equations contain the results described in [1819] as special cases; from what he says in the abstract of [1819], it seems likely that in that memoir he was concerned only with the case $\varphi(x) = x$.

Equation (3.31) also holds when multiple zeros of $F(x)$ are present, provided that each zero is counted a number of times equal to its multiplicity; there is no evidence that Cauchy was aware of the fact at the time, since it was only in the 1825 memoir that he began to cope efficiently with multiple roots (Chapter 4 below).

We note also that in an earlier section of the 1823 memoir Cauchy gives[40] some quite different formulae for the roots of an equation; these were much more complicated than the ones we have discussed in this section, being based on the Fourier integral formula and involving multiple integrals.

3.12. During the period we are concerned with Cauchy obtained several results about power series.

Chapter IX of the *Analyse algébrique* is devoted to infinite series with

---

[40] [1823a], pp. 541–3; *Oeuvres* (2)1, 303–7.

complex terms, power series with complex coefficients being dealt with in Section 2. His main result here[41] is that the series

$$a_0 + a_1 z + a_2 z^2 + \cdots$$

is convergent for $|z| < 1/A$ and divergent for $|z| > 1/A$, where (in modern notation)

$$A = \limsup_{n \to \infty} |a_n|^{1/n}.$$

This result on the radius of convergence seems to have been almost forgotten until it was rediscovered by Hadamard in 1892 (although it seems to have been known to Riemann).

The note [1822a] begins with a discussion of the ways in which power series were used. He makes the point that even those mathematicians who do not adopt Lagrange's point of view about the foundations of the calculus nevertheless use such expansions for a number of purposes, including (i) the determination of the number of arbitrary constants or arbitrary functions appearing in the general integral of a differential equation, (ii) the calculation of such integrals, and (iii) methods of distinguishing between singular and other integrals. In all these applications it is taken for granted that a function is completely determined by its Maclaurin or Taylor series.

At this point he drops his bombshell, showing that the Maclaurin series of the function e^{-1/x^2} vanishes identically, and consequently does not represent the function; he also gives some other examples of the same phenomenon. He draws the moral that a function may be replaced by its Taylor or Maclaurin series only if the series is convergent and, in addition, has the function as its sum. In other cases, he says, one must work with a partial sum of the series, together with a remainder that can be evaluated or estimated; this must be borne in mind in considering problems about maxima and minima, indeterminate forms (limits of the type 0/0) and similar matters.

In the remainder of the note he illustrates his point by discussing some special differential equations, including an initial-value problem with a non-unique solution, and draws the conclusion that one cannot assume *a priori* that the solution of a differential equation can always be expanded in a power series.

The example e^{-1/x^2} is also mentioned briefly at the end of Lesson 38 of the *Calcul infinitésimal* [1823b]. It was of course a severe blow to Lagrange's programme of basing the calculus entirely on power-series

[41] [1821a], pp. 286–7; *Oeuvres* (2)**3**, 239–40.

expansions. It is also the earliest example of an infinitely differentiable function of a real variable that is not analytic on the whole real line; it thus opened the way towards the eventual recognition of the fact that the analytic functions form a relatively small subclass of the whole body of continuous functions.

In Lesson 40 of the *Calcul infinitésimal* [1823b] Cauchy introduces a novel argument concerning power series. His result states that if the functions $f(x, z)$ and

$$F(x) = \int_{z_0}^{Z} f(x, z)\,dz \qquad (3.33)$$

can be expanded as power series in x for $z_0 \leqslant z \leqslant Z$ and for real values of x with $-r \leqslant x \leqslant r$, and if we continue to denote the sums of these series by $f(x, z)$ and $F(x)$ when x is complex, then (3.33) will still hold for complex values of x when $|x| < r$.

To illustrate the usefulness of this result Cauchy starts from the equation

$$\int_{-\infty}^{\infty} e^{-(z+x)^2}\,dz = \sqrt{\pi},$$

which can be rewritten in the equivalent form

$$\int_{0}^{\infty} e^{-z^2} \cosh 2zx\,dz = \tfrac{1}{2}\sqrt{\pi}\,e^{+x^2}.$$

Since this holds for all real x, it follows from his general result that it is true also for all complex x. Replacing x by ix, he deduces the equation

$$\int_{0}^{\infty} e^{-z^2} \cos 2zx\,dz = \tfrac{1}{2}\sqrt{\pi}\,e^{-x^2}.$$

He is assuming here that he can integrate term by term over an infinite interval, which would require more justification than he provides.

It is interesting to note that he is here using expansibility in power series as a hypothesis, so that the notion of analyticity in this sense is beginning to play a role in Cauchy's framework for analysis. He is still far, however, from any recognition of the intimate relations between expansibility in power series and the property of being holomorphic.

3.13. We now summarise briefly, and approximately in chronological order, Cauchy's main improvements to the theory during the period covered by the present chapter.

Probably from 1817 onwards, and certainly from 1819, Cauchy abandoned the exclusive use of real integrands; the admission of complex integrands throughout brought about a notable simplification of the theory and widened the class of cases to which it could be applied (Sections 3.2 and 3.3).

In 1817 and 1820 he gave proofs of the fundamental theorem of algebra for real polynomials, and in 1821 he extended the result to complex polynomials (Section 3.11); in this he had been partially anticipated by Argand in 1815.

In 1821 he established the existence of the radius of convergence of a power series, and in 1822 he discovered the existence of functions whose Maclaurin series was convergent, but not to the function itself (Section 3.12).

Improved reformulations of the basic results of the 1814 memoir appeared in 1822 and 1823 (Section 3.5).

In 1822 he obtained results equivalent to what were later called the 'Cauchy formulae' for the values of functions at points inside a circular contour, and for their derivatives at the centre of the circle, in terms of integrals round the circle (Section 3.7). In the same year he developed the theory of principal values for the integrals of functions having an infinity in the range of integration (Section 3.8).

In 1823 he improved the proof of the basic formula for his correction term (equivalent to the residue theorem for a rectangle when the function has only simple poles) (Section 3.6). Later in the same year he gave, without proof, an expression for the correction term associated with a multiple pole (Section 3.10). Evidence also began to emerge that his standard conditions for the behaviour of functions at infinity were inadequate, but he was to remain puzzled by the discrepancy for some time to come (Section 3.9; cf. also Section 4.11 below).

In 1825 he completed the footnotes to be inserted in the published version of the 1814 memoir, mainly indicating the improvements due to the admission of complex integrands (Section 3.2); the memoir was ultimately published in 1827.

4

The 1825 memoir and associated articles

4.1. On 28 February 1825 Cauchy communicated to the Académie des Sciences a memoir entitled 'Mémoire sur les intégrales définies prises entre des limites imaginaires'. He published an abstract of it in the *Bulletin de Férussac* [1825a] in April of the same year. The complete memoir was issued as a separate pamphlet in August [1825b]. It was republished by Darboux in the *Bulletin des Sciences Mathématiques* in 1874–5, and a German translation of it appeared in the series *Ostwalds Klassiker der Exakten Wissenschaften* in 1900, with notes and commentary by P. Stäckel.

We shall usually refer to [1825b] as the 1825 memoir; it constituted an important advance in Cauchy's development of complex function theory.

In the present chapter we shall also include some account of a paper by Cauchy, published in two parts, [1825c] and [1826f], in Gergonne's *Annales de Mathématiques*; it can most conveniently be regarded as a sequel to the 1825 memoir.

4.2. In his introduction to the 1825 memoir Cauchy recalls his use of singular integrals in the 1814 memoir and his results on principal values in [1822b], [1823a] and in the *Analyse infinitésimale* [1823b], remarking that these together had enabled him to establish numerous general results on the evaluation or transformation of definite integrals. He now proposes to apply the same principles to integrals between imaginary limits.

He recalls Laplace's use of such integrals, and mentions that Brisson had recently used them to obtain expansions of functions in exponential series. Brisson's work has remained unpublished; he submitted a memoir to the Academy on 17 November 1823, some aspects of which were outlined by Cauchy in his referee's report [1825d] dated 13 June 1825, and some of Brisson's work in this direction may have been contained in it.

Cauchy goes on to acknowledge some contributions from Ostrogradskiĭ, who had found new proofs for some of his results and had generalised some of the formulae of [1823a]. A conjecture about the nature of Ostrogradskiĭ's contributions is described later (Section 4.6 below).

Another paper of interest in this connection is Poisson [1820]. In the second part of this paper (pp. 318ff) he considers integrals of functions that become infinite between the limits of integration. He remarks that, according to the rules usually accepted, one should have

$$\int_{-1}^{1} \frac{dx}{x^2} = \left[-\frac{1}{x} \right]_{-1}^{1} = -2,$$

a negative result, although dx/x^2 is positive throughout the interval. Again, we should have

$$\int_{-1}^{1} \frac{dx}{x} = [\log x]_{-1}^{1} = -\log(-1),$$

which has the infinity of values $(2n + 1)\pi i$; how, he asks, can the sum of the real elements dx/x have several values, all of them imaginary? He concludes that the usual proof that if $dF(x) = f(x)\,dx$, then $F(b) - F(a)$ is the sum of the values of $f(x)\,dx$ when x goes from a to b by infinitely small steps, ceases to be valid if $f(x)$ becomes infinite between a and b.

Poisson goes on to suggest that the validity of the general result can be restored by making an appropriate change of variable. In the above examples, one could put

$$x = -(\cos \zeta + i \sin \zeta),$$

and integrate from $\zeta = 0$ to $\zeta = (2n + 1)\pi$, where n is an integer. Since x no longer passes through 0, the integrand no longer has an infinity. One then obtains

$$\int_{-1}^{1} \frac{dx}{x} = -i \int_{0}^{(2n+1)\pi} d\zeta = -(2n + 1)\pi i,$$

the result we had earlier. He then evaluates

$$\int_{-1}^{1} \frac{dx}{x^m}$$

in the same way for positive integral values of m, and so obtains an explanation for the negative result found for $m = 2$.

Another example that Poisson considers is the integral

$$y = \int_{-\infty}^{\infty} \frac{\cos ax \cdot dx}{b^2 + x^2}.$$

He makes the substitution

$$x = t + k\mathrm{i},$$

where k is a real constant, and shows that if a and b are both positive, then

$$y = \frac{\pi}{b}\,\mathrm{e}^{-ab} \qquad (0 < k < b),$$

but

$$y = \frac{\pi}{2b}(\mathrm{e}^{-ab} - \mathrm{e}^{ab}) \qquad (k > b).$$

We have noted earlier (Section 3.8 above) that Cauchy referred to this paper of Poisson's in his [1822b] and [1823a]; we there discussed the relation between some of Poisson's results and Cauchy's concept of principal-value integrals.

In another paper by Poisson [1823a], he recognises (p. 460) that if one makes the substitution $x' = x + y\mathrm{i}$ in an integral of the form

$$\int_{-\infty}^{\infty} g(x)\,\mathrm{d}x,$$

then the integral becomes

$$\int_{-\infty}^{\infty} g(x')\,\mathrm{d}x',$$

where the variable x' now runs through imaginary values instead of real ones, although the limits of integration are still denoted by $(-\infty, \infty)$.

In these examples Poisson has clearly grasped that one has to think in terms of integration along a particular path, rather than considering only what the limits of integration may be.

4.3. In Section 2 of the 1825 memoir Cauchy presents his definition for integrals taken between complex limits. He recalls his own definition in Lesson 21 of the *Calcul infinitésimal* for the definite integral of a function of a real variable; to extend it to functions of a complex variable, he initially defines

$$\int_{x_0 + y_0\mathrm{i}}^{X + Y\mathrm{i}} f(z)\,\mathrm{d}z$$

as the limit (or one of the limits) of sums of the form

$$[(x_1 - x_0) + (y_1 - y_0)\mathrm{i}]f(x_0 + y_0\mathrm{i}) + [(x_2 - x_1) + (y_2 - y_1)\mathrm{i}]f(x_1 + y_1\mathrm{i})$$
$$+ \cdots + [(X - x_{n-1}) + (Y - y_{n-1})\mathrm{i}]f(x_{n-1} + y_{n-1}\mathrm{i}),$$

where the sequences $(x_0, x_1, \ldots, x_{n-1}, X)$ and $(y_0, y_1, \ldots, y_{n-1}, Y)$ are both monotonic (either increasing or decreasing) and the differences $x_k - x_{k-1}$ and $y_k - y_{k-1}$ converge to zero as n increases indefinitely.

We note that at this stage Cauchy is not constructing the integral along a definite path, but merely has successive sequences of points running from $x_0 + y_0 i$ to $X + Y i$; presumably this is one reason why he uses the phrase 'one of the limits'. The requirement that the sequences should be monotonic seems to have been carried over almost automatically from his definition for functions of a real variable.

He immediately suggests a procedure for constructing such sequences. Let us write

$$x = \varphi(t), \quad y = \chi(t),$$

where $\varphi(t)$ and $\chi(t)$ are continuous and increasing from $t = t_0$ to $t = T$, and satisfy the conditions

$$\varphi(t_0) = x_0, \quad \chi(t_0) = y_0,$$

$$\varphi(T) = X, \quad \chi(T) = Y,$$

and let

$$x_k = \varphi(t_k), \quad y_k = \chi(t_k) \qquad (k = 1, 2, \ldots, n - 1),$$

where $(t_0, t_1, \ldots, t_{n-1}, T)$ is a monotonic sequence. Assuming tacitly that $\varphi'(t)$ and $\chi'(t)$ exist and are continuous, he argues that we have approximately

$$x_k - x_{k-1} = (t_k - t_{k-1})\varphi'(t_{k-1}),$$

$$y_k - y_{k-1} = (t_k - t_{k-1})\chi'(t_{k-1}).$$

The integral

$$A + Bi = \int_{x_0 + y_0 i}^{X + Y i} f(z)\, dz$$

must therefore be nearly equal to the expression

$$(t_1 - t_0)[\varphi'(t_0) + i\chi'(t_0)]f[\varphi(t_0) + i\chi(t_0)]$$

$$+ (t_2 - t_1)[\varphi'(t_1) + i\chi'(t_1)]f[\varphi(t_1) + i\chi(t_1)]$$

$$+ \cdots + (T - t_{n-1})[\varphi'(t_{n-1}) + i\chi'(t_{n-1})]f[\varphi(t_{n-1}) + i\chi(t_{n-1})],$$

which is an approximating sum for the integral

$$\int_{t_0}^{T} [\varphi'(t) + i\chi'(t)]f[\varphi(t) + i\chi(t)]\, dt.$$

Hence, he says, we must have

$$A + Bi = \int_{t_0}^{T} [\varphi'(t) + i\chi'(t)]f[\varphi(t) + i\chi(t)]\, dt;$$

more briefly, if we write

$$\varphi'(t) = x', \quad \chi'(t) = y',$$

we have

$$A + Bi = \int_{t_0}^{T} (x' + y'i)f(x + yi)\,dt.$$

We see that he has here reduced an integral between complex limits to the integral of a complex-valued function between real limits. His argument does not meet present-day standards of rigour, but it could easily be made rigorous under appropriate hypotheses.

In the more geometrical language that he was to adopt later in the present memoir (Section 4.8 below), he has constructed the integral of $f(z)$ along the path in the plane given by the parametric representation

$$x = \varphi(t), \quad y = \chi(t) \qquad (t_0 \leqslant t \leqslant T).$$

4.4. In Section 3 of the 1825 memoir, Cauchy explicitly assumes that $f(x + yi)$ is 'finite and continuous' for values of x and y such that $x_0 \leqslant x \leqslant X$ and $y_0 \leqslant y \leqslant Y$; he takes it for granted that under these conditions $f'(x + yi)$ will also exist and be continuous. His announced aim is to prove that the value of the integral

$$\int_{x_0 + y_0 i}^{X + Y i} f(z)\,dz$$

is independent of the choice of the functions $x = \varphi(t)$ and $y = \chi(t)$. In geometrical language, his result is to be that the value of the integral is the same for all monotonic paths from $x_0 + y_0 i$ to $X + Y i$.

He argues as follows. Let us replace x and y by $x + \varepsilon u$ and $y + \varepsilon v$, where $u(t)$ and $v(t)$ both vanish when $t = t_0$ and when $t = T$, and ε is infinitely small (i.e. ε is a variable that will ultimately tend to 0). The integral will then receive a corresponding increment, which can be expanded in a series of powers of ε; the coefficient of ε in this series will be

$$\int_{t_0}^{T} [(u + vi)(x' + y'i)f'(x + yi) + (u' + v'i)f(x + yi)]\,dt.$$

Integrating by parts, we see that

$$\int_{t_0}^{T} (u' + v'i)f(x + yi)\,dt = -\int_{t_0}^{T} (u + vi)f'(x + yi)(x' + y'i)\,dt;$$

the coefficient of ε therefore vanishes, and so the increment of the integral must be of the second or higher order in ε. From this he concludes that if we give x and y successive increments of the first order whose sum is a finite non-zero increment, then the corresponding increment of the integral will be of the first order; in other words, it must vanish.

Cauchy then rephrases the argument in the language of the calculus of variations. He writes

$$u = \delta x, \quad v = \delta y$$

(more correctly, he should have written $\varepsilon u = \delta x$, $\varepsilon v = \delta y$), so that the increment of the integral becomes

$$\delta \int_{t_0}^{T} (x' + y'\mathrm{i}) f(x + y\mathrm{i}) \, \mathrm{d}t$$

$$= \int_{t_0}^{T} [(x' + y'\mathrm{i}) \delta f(x + y\mathrm{i}) + f(x + y\mathrm{i}) \delta(x' + y'\mathrm{i})] \, \mathrm{d}t$$

$$= 0.$$

He finally remarks that this result could have been foreseen, since the expression

$$f(x + y\mathrm{i})(\mathrm{d}x + \mathrm{i} \, \mathrm{d}y)$$

is an exact differential; presumably he means here that if we write

$$U \, \mathrm{d}x + V \, \mathrm{d}y = f(x + y\mathrm{i})(\mathrm{d}x + \mathrm{i} \, \mathrm{d}y),$$

then we have

$$\frac{\partial U}{\partial y} = \frac{\partial V}{\partial x} = \mathrm{i} f'(x + y\mathrm{i}).$$

It is not clear whether Cauchy regarded this argument as a proof.

We note that, in modern terms, Cauchy's argument involves the continuous deformation of one curve into another; as it stands, it is some way from being rigorous by present-day standards. Various writers have constructed more rigorous versions of it; the earliest seems to be that of M. Falk [1883], whose proof brings out more clearly the homotopy flavour of the argument.

Falk supposes that the function $F(z)$ is single-valued and continuous and has a continuous derivative in a domain bounded by a simple closed contour and that, for each value of ρ with $0 \leqslant \rho \leqslant 1$, the equation

$$z = \varphi(t) + \rho \psi(t) \qquad (\alpha \leqslant t \leqslant \beta)$$

represents a curve in the interior of the domain such that $z = z_0$ when $t = \alpha$ and $z = z_1$ when $t = \beta$, so that $\psi(\alpha) = \psi(\beta) = 0$; he also assumes that $\varphi(t)$ and $\psi(t)$ are continuous and have continuous derivatives. Along the curve with parameter ρ, we have

$$\int_{z_0}^{z_1} F(z)\, dz = \int_{\alpha}^{\beta} F[\varphi(t) + \rho\psi(t)][\psi'(t) + \rho\psi'(t)]\, dt.$$

It follows that

$$\frac{d}{d\rho}\int_{z_0}^{z_1} F(z)\, dz = \int_{\alpha}^{\beta} F[\psi(t) + \rho\psi(t)]\psi'(t)\, dt$$

$$+ \int_{\alpha}^{\beta} F'[\varphi(t) + \rho\psi(t)]\psi(t)[\varphi'(t) + \rho\psi'(t)]\, dt$$

$$= [F(\varphi(t) + \rho\psi(t))\psi(t)]_{\alpha}^{\beta}$$

$$= 0.$$

The value of the integral is therefore independent of ρ.

Similar proofs have been given by W. F. Eberlein [1979] and G. C. Berresford [1981].

4.5. In Section 4 of the 1825 memoir Cauchy takes up the case where $f(z)$ has an infinity at $z = a + bi$; he boldly supposes that $a + bi$ actually lies on the path of integration for the integral

$$A + Bi = \int_{t_0}^{T} (x' + y'i)f(x + yi)\, dt,$$

and writes τ for the corresponding value of the parameter t. He also assumes that

$$[(x - a) + (y - b)i]f(x + yi)$$

has a limit, which we denote by f_0 (a slight change from Cauchy's notation), when x and y converge to a and b respectively; in modern terms, he is assuming that $a + bi$ is a simple pole of $f(z)$ with residue f_0. He then says that, if ε is infinitely small, then we have, approximately.

$$f_0 = \varepsilon f(a + bi + \varepsilon).$$

He gives $x(t)$ and $y(t)$ the infinitely small increments $\varepsilon u(t)$ and $\varepsilon v(t)$, where $u(t) = v(t) = 0$ when $t = t_0$ and when $t = T$. Denoting the original value of the integral

$$\int_{x_0 + y_0 i}^{X + Yi} f(z)\, dz$$

(taken as a principal value, as he later remarks) by $A + Bi$ and its value along the perturbed path by $A' + B'i$, he has

$(A' + B'i) - (A + Bi)$

$$= \int_{t_0}^{T} [(x' + \varepsilon u') + (y' + \varepsilon v')i] f[(x + \varepsilon u) + (y + \varepsilon v)i] \, dt$$

$$- \int_{t_0}^{T} (x' + y'i) f(x + yi) \, dt. \tag{4.1}$$

He then remarks that the difference between the two integrals on the right-hand side of (4.1) will (by what he has already proved) be practically zero except when x is near a and y is near b, i.e. when t is near τ. He therefore makes the substitution

$$t = \tau + \varepsilon w,$$

taking w to range from $-\varepsilon^{-1/2}$ to $+\varepsilon^{-1/2}$, thus neglecting the contributions from the parts of the paths remote from the singularity; he writes

$$x'(\tau) = \alpha, \quad y'(\tau) = \beta, \quad u(\tau) = \gamma, \quad v(\tau) = \delta$$

and

$$\lambda + \mu i = \frac{\gamma + \delta i}{\alpha + \beta i},$$

so that we have, in particular,

$$\mu = \frac{\alpha \delta - \beta \gamma}{\alpha^2 + \beta^2}.$$

He now argues that, since εw is small, we have approximately

$$(x' + y'i) f(x + yi) = \frac{(x' + y'i) f_0}{(x - a) + (y - b)i}$$

$$= \frac{(x' + y'i) f_0}{(x' + y'i)(t - \tau)}$$

$$= \frac{f_0}{\varepsilon w};$$

similarly

$$[(x' + \varepsilon u') + (y' + \varepsilon v')i] f[(x + \varepsilon u) + (y + \varepsilon v)i]$$

is nearly equal to

$$\frac{f_0}{\varepsilon(w + \lambda + \mu i)}.$$

(The argument for this last result seems to be that we have approximately

$$f[(x + \varepsilon u) + (y + \varepsilon v)] = \frac{f_0}{(x - a + \varepsilon u) + (y - b + \varepsilon v)\mathrm{i}}$$

$$= \frac{f_0}{(x' + y'\mathrm{i})(t - \tau) + \varepsilon(u + v\mathrm{i})}$$

$$= \frac{f_0}{\varepsilon} \frac{1}{w(x' + y'\mathrm{i}) + (u + v\mathrm{i})}$$

$$= \frac{f_0}{\varepsilon} \frac{1}{w(\alpha + \beta\mathrm{i}) + (\gamma + \delta\mathrm{i})}$$

$$= \frac{f_0}{\varepsilon} \frac{1}{\alpha + \beta\mathrm{i}} \frac{1}{w + (\lambda + \mu\mathrm{i})};$$

the factor $(x' + \varepsilon u') + (y' + \varepsilon v')\mathrm{i}$ can be replaced by $x' + y'\mathrm{i}$ and hence by $\alpha + \beta\mathrm{i}$, the contribution from the neglected terms being of a smaller order of magnitude.)

We therefore have, to the required order of approximation,

$$(A' + B'\mathrm{i}) - (A + B\mathrm{i}) = f_0 \int_{-\varepsilon^{-1/2}}^{\varepsilon^{-1/2}} \frac{\mathrm{d}w}{w + (\lambda + \mu\mathrm{i})} - f_0 \int_{-\varepsilon^{-1/2}}^{\varepsilon^{-1/2}} \frac{\mathrm{d}w}{w}$$

$$= f_0 \int_{-\varepsilon^{-1/2}}^{\varepsilon^{-1/2}} \frac{\mathrm{d}w}{w + (\lambda + \mu\mathrm{i})},$$

since the second term on the right is a vanishing principal-value integral. What remains is then, approximately,

$$f_0 \int_{-\infty}^{\infty} \frac{\mathrm{d}w}{w + (\lambda + \mu\mathrm{i})} = -f_0\mathrm{i} \int_{-\infty}^{\infty} \frac{\mu\,\mathrm{d}w}{(w + \lambda)^2 + \mu^2}$$

$$= \mp \pi f_0 \mathrm{i}$$

according as $\mu > 0$ or $\mu < 0$, another vanishing principal-value integral having been discarded. We thus obtain finally

$$(A' + B'\mathrm{i}) - (A + B\mathrm{i}) = \mp \pi f_0 \mathrm{i} \tag{4.2}$$

according as $\alpha\delta - \beta\gamma \gtrless 0$ or, equivalently

$$x'(t)v(t) - y'(t)u(t) \gtrless 0,$$

or, more briefly,

$$x'\delta y - y'\delta x \gtrless 0.$$

Cauchy concludes this discussion by remarking that, if we give x and y increments of the opposite sign to those just used and denote the value of the integral so obtained by $A'' + B''\mathrm{i}$, then we shall have

$$(A'' + B''\mathrm{i}) - (A + B\mathrm{i}) = \pm \pi F_0 \mathrm{i},$$

whence it follows that

$$(A'' + B''\mathrm{i}) - (A' + B'\mathrm{i}) = \pm 2\pi f_0 \mathrm{i}. \tag{4.3}$$

The two integrals compared in (4.3) will not, in general, have to be interpreted as principal values.

In Section 5 of the memoir Cauchy gives an alternative derivation of the same result; here he introduces the idea of splitting the integrand $f(z)$ into two parts, of which one is a simple rational expression having the same behaviour as $f(z)$ near the singularity and the other is a well-behaved function. He writes

$$f(z) = \frac{f_0}{z - a - b\mathrm{i}} + \varpi(z),$$

so that $\varpi(z)$ does not become infinite at $z = a + b\mathrm{i}$. The contribution of $\varpi(z)$ to the difference $(A' + B'\mathrm{i}) - (A + B\mathrm{i})$ then vanishes, since both its terms are, by what has already been proved (Section 4.4 above) equal to the unique value of the integral

$$\int_{x_0 + y_0\mathrm{i}}^{X + Y\mathrm{i}} \varpi(z)\,\mathrm{d}z.$$

The contribution of the remaining expression to the difference is then

$$f_0 \int_{t_0}^{T} \left\{ \frac{(x' + \varepsilon u') + (y' + \varepsilon v')\mathrm{i}}{(x - a + \varepsilon u) + (y - b + \varepsilon v)\mathrm{i}} - \frac{x' + y'\mathrm{i}}{(x - a) + (y - b)\mathrm{i}} \right\} \mathrm{d}t$$

$$= f_0 [\log((x - a + \varepsilon u) + (y - b + \varepsilon v)\mathrm{i}) - \log((x - a) + (y - b)\mathrm{i})]_{t_0}^{T}$$

$$= f_0 [\tfrac{1}{2}\log((x - a + \varepsilon u)^2 + (y - b + \varepsilon v)^2) - \tfrac{1}{2}\log((x - a)^2 + (y - b)^2)]_{t_0}^{T}$$

$$+ \; \mathrm{i}f_0 \left[\arctan \frac{y - b + \varepsilon v}{x - a + \varepsilon u} - \arctan \frac{y - b}{x - a} \right]_{t_0}^{T}.$$

Since we are taking the principal value of the integral, and the functions $u(t)$ and $v(t)$ vanish at $t = t_0$ and $t = T$, the real part of the coefficient of f_0 reduces to

$$\tfrac{1}{2} \left[\log \frac{(x - a + \varepsilon u)^2 + (y - b + \varepsilon v)^2}{(x - a)^2 + (y - b)^2} \right]_{t_0}^{T} = 0.$$

The imaginary part can be written in the form

$$\left[\arctan \frac{\varepsilon((x - a)v - (y - b)u)}{(x - a)(x - a + \varepsilon u) + (y - b)(y - b + \varepsilon v)} \right]_{t_0}^{T};$$

this is the total change in the expression in square brackets as t goes from t_0 to T. To evaluate it correctly, one has to examine separately the change from t_0 to the value τ where the singularity occurs and that from τ to T; each of these contributes $\mp\frac{1}{2}\pi$ to the imaginary part. We thus obtain

$$(A' + B'\mathrm{i}) - (A + B\mathrm{i}) = \mp\pi_0 f\mathrm{i},$$

the same result as before.

4.6. In Section 6 of the 1825 memoir Cauchy takes up the problem of discovering what happens when the equation $1/f(z) = 0$ has a root of order $m > 1$ at $z = a + b\mathrm{i}$. We recall (Section 3.10 above) that in a footnote to [1823a] he had given an expression equivalent to the value of

$$\frac{2\pi\mathrm{i}}{(m-1)!}\frac{\mathrm{d}^{m-1}[k^m f(x_0 + k)]}{\mathrm{d}k^{m-1}}$$

at $k = 0$ for the contribution of such a root, but without any proof. He now embarks on a proof (indeed two proofs) for this case.

He begins by writing (modulo a slight change of notation)

$$f_0(z) = (z - a - b\mathrm{i})^m f(z),$$

so that $f_0(z)$ does not have an infinity at $z = a + b\mathrm{i}$. This time he starts with complementary paths not passing through the singularity, writing (as before) $A' + B'\mathrm{i}$ for the value of the integral

$$\int_{x_0 + y_0\mathrm{i}}^{X + Y\mathrm{i}} f(z)\,\mathrm{d}z$$

when $x(t)$ and $y(t)$ receive increments $\varepsilon u(t)$ and $\varepsilon v(t)$ respectively and $A'' + B''\mathrm{i}$ for its value when the increments are $-\varepsilon u(t)$ and $-\varepsilon v(t)$. We then have

$$(A'' + B''\mathrm{i}) - (A' + B'\mathrm{i})$$

$$= \int_{t_0}^{T} [(x' - \varepsilon u') + (y' - \varepsilon v')\mathrm{i}]\frac{f_0[(x - \varepsilon u) + (y - \varepsilon v)\mathrm{i}]}{[(x - a - \varepsilon u) + (y - b - \varepsilon v)\mathrm{i}]^m}\,\mathrm{d}t$$

$$- \int_{t_0}^{T} [(x' + \varepsilon u') + (y' + \varepsilon v')\mathrm{i}]\frac{f_0[(x + \varepsilon u) + (y + \varepsilon v)\mathrm{i}]}{[(x - a + \varepsilon u) + (y - b + \varepsilon v)\mathrm{i}]^m}\,\mathrm{d}t.$$

$$(4.4)$$

Tacitly assuming, as usual, that $f_0(z)$ has as many derivatives as are needed, he defines $\varpi(z)$ by the equation

$$f_0(z) = f_0(a + b\mathrm{i}) + \frac{f_0'(a + b\mathrm{i})}{1!}(z - a - b\mathrm{i}) + \cdots$$

$$+ \frac{f_0^{(m-1)}(a + b\mathrm{i})}{(m-1)!}(z - a - b\mathrm{i})^{m-1} + (z - a - b\mathrm{i})^m \varpi(z);$$

by hypothesis, $\varpi(z)$ is well-behaved near $z = a + b\mathrm{i}$.

Substituting this expression in (4.4), and noting that the contribution from the term containing $\varpi(z)$ will vanish, he obtains

$$(A'' + B''\mathrm{i}) - (A' + B'\mathrm{i}) = s_0 f_0(a + b\mathrm{i}) + \frac{s_1}{1!} f_0'(a + b\mathrm{i}) + \cdots$$

$$+ \frac{s_{m-2}}{(m-2)!} f_0^{(m-2)}(a + b\mathrm{i})$$

$$+ \frac{s_{m-1}}{(m-1)!} f_0^{(m-1)}(a + b\mathrm{i}),$$

where

$$s_n = \int_{t_0}^{T} \left\{ \frac{(x' - \varepsilon u') + (y' - \varepsilon v')\mathrm{i}}{[(x - a - \varepsilon u) + (y - b - \varepsilon v)\mathrm{i}]^{m-n}} \right.$$

$$\left. - \frac{(x' + \varepsilon u') + (y' + \varepsilon v')\mathrm{i}}{[(x - a + \varepsilon u) + (y - b + \varepsilon v)\mathrm{i}]^{m-n}} \right\} dt \qquad (n = 0, 1, \ldots, m - 1).$$

These expressions can be evaluated directly; it turns out that s_n vanishes except when $n = m - 1$, in which case we have $s_{m-1} = \pm 2\pi\mathrm{i}$. It follows that

$$(A'' + B''\mathrm{i}) - (A' + B'\mathrm{i}) = \pm 2\pi\mathrm{i} \frac{f_0^{(m-1)}(a + b\mathrm{i})}{(m-1)!};$$

the right-hand side can also be expressed as

$$\pm 2\pi\mathrm{i} \lim_{\varepsilon \to 0} \frac{1}{(m-1)!} \frac{\mathrm{d}^{m-1}}{\mathrm{d}\varepsilon^{m-1}} [\varepsilon^m f(a + b\mathrm{i}) + \varepsilon],$$

thus equation (4.3) of Section 4.5 is still valid, except that we now have to take

$$f_0 = \frac{f_0^{(m-1)}(a + b\mathrm{i})}{(m-1)!},$$

the ambiguous $\pm$ sign being determined as before.

We note that he has here used again the device of separating off the singular part of the function, introduced in his second proof for the case of a simple pole (Section 4.5 above).

Cauchy then gives an alternative proof that $s_n = 0$ for $n < m - 1$ and $s_{m-1} = \pm 2\pi\mathrm{i}$; in this he uses an approximation argument similar to that in the first proof described in Section 4.5 above. We omit the rather tedious details. It is odd that he should have included this argument, since, as we have seen, the expressions for the s_n can be evaluated explicitly, and no approximation is needed. A plausible speculation would be that the second proof for $m = 1$ and the first proof for $m > 1$ were inserted at a late stage in the preparation of this part of the memoir. There may well have been an earlier version in which the approximation argument was the only one given for $m = 1$, and there would have been an analogous argument for $m > 1$; in the version we have, the latter argument was recast, being applied only to the integral expressions for the s_n, where it is actually completely superfluous.

The device of splitting the function into a simple rational expression for its singular part and a well-behaved part appears here for the first time in Cauchy's publications, but a similar device was used by Ostrogradskiĭ in an unpublished paper, written during his stay in Paris and dated 24 July 1824, and described by Yushkevich [1965], and it reappears in a later paper by Ostrogradskiĭ [1828]. As we mentioned in Section 4.2 above. Cauchy acknowledges in his introduction to the 1825 memoir that Ostrogradskiĭ had given him new proofs for some of his results; it thus seems likely that this device was contributed by Ostrogradskiĭ. Yushkevich suggests that Ostrogradskiĭ's new technique first appeared in the initial paper [1826a] of Cauchy's series on the theory of residues (Section 5.3 below), but the argument here is of a similar pattern. The technique of splitting the function appears again in [1826c] (Section 5.4 below).

In Section 7 of the memoir Cauchy investigates what happens when the unperturbed path of integration actually passes through a root $a + bi$ of order $m > 1$ of the equation $1/f(z) = 0$. If the integrals along this path and the perturbed path are $A + Bi$ and $A' + B'i$, respectively, it turns out, as one would expect, that $(A' + B'i) - (A + Bi)$ is in general infinite, even when $A + Bi$ is interpreted as a principal-value integral. An exception occurs when

$$f_0^{(m-2)}(a + bi) = f_0^{(m-4)}(a + bi) = \cdots = 0;$$

in this special case he finds that

$$(A' + B'i) - (A + Bi) = \mp \frac{\pi i f_0^{(m-1)}(a + bi)}{(m-1)!}.$$

One gets the impression from this passage that Cauchy looked upon his creation of the notion of principal value as a contribution of major importance to the integral calculus, so much so that he introduced such expressions in every case where they might be meaningful.

4.7. In Section 8 of the 1825 memoir Cauchy makes the obvious remark that if there are several values z_1, z_2, $\ldots$ lying between the paths of integration at which $f(z)$ becomes infinite, and if the corresponding values of the constant f_0 are $f_1, f_2, \ldots$, then we shall have

$$(A'' + B''i) - (A' + B'i) = \pm 2\pi i f_1 \pm 2\pi i f_2 \pm \cdots.$$

He also discusses

$$(A' + B'\mathrm{i}) - (A + B\mathrm{i})$$

in this situation, remarking that it will in general be infinite unless the special conditions mentioned in Section 7 (Section 4.6 above) hold at each singularity. In particular, it will usually be infinite if each of the relevant zeros of $1/f(z)$ is of even order; he even goes to the trouble of working through in detail, as an illustrative example, the integral

$$\int_{-1-i}^{1+i} \frac{\mathrm{d}z}{z^2(1 + z^2)},$$

taken along the path given by $z = t + t\mathrm{i} \, (-1 \leqslant t \leqslant 1)$.

One's impression here is that Cauchy had a feeling of disappointment that the use of principal values was not of such general applicability as he had originally hoped.

4.8. In Section 9 of the 1825 memoir Cauchy suddenly introduces, without any prior warning (apart from a brief mention in the preliminary abstract [1825a]), some geometrical language in the description of his results. Before this time he had carefully avoided the use of any such language; from now on, however, he uses it, occasionally to begin with, and ultimately more and more freely. As Belhoste [1991] remarks (p. 215), Cauchy was strongly opposed to those in the Ecole Polytechnique who maintained that analysis should be presented in a 'simple and straightforward' manner, with its possible applications to engineering problems being kept continually in mind. Cauchy felt that this view tended to encourage appeals to intuition and the use of geometrical language; as a disciple of Lagrange, he mistrusted such procedures, which, he believed, would tend to disguise the general applicability of analytic results. At this point, however, he seems to find that he can express some of his ideas more concisely by admitting geometrical terminology.

He begins with the remark that if one eliminates t from the equations (Section 4.3 above)

$$x = \varphi(t), \qquad y = \chi(t),$$

one obtains an equation of the form

$$F(x, y) = 0. \tag{4.5}$$

If (x, y) are regarded as rectangular coordinates in the plane, then (4.5) will represent a curve joining the points (x_0, y_0) and (X, Y). Under the conditions imposed on $\varphi(t)$ and $\chi(t)$, y will be uniquely determined as a function of x and the curve will lie in the rectangle formed by the lines

$$x = x_0, \ x = X, \ y = y_0, \ y = Y.$$

To each such curve will correspond a value of the integral

$$\int_{x_0 + y_0 i}^{X + Y i} f(z) \, dz.$$

He goes on to say that the whole path may be made up of portions defined by different functions, provided only that the functions defining the separate portions satisfy his conditions and that the whole path begins at (x_0, y_0) and ends at (X, Y). In particular, a portion of the path may be a straight segment parallel to one of the coordinate axes. He writes down explicit formulae for the contribution of a single portion to the whole integral, whether the portion is a straight segment or is given by an equation of the form $y = \psi(x)$ or $x = \psi(y)$. We see that he is now requiring each portion of the path to be monotonic, but not the path as a whole.

We note that he is not using the full force of Argand's geometrical representation for complex numbers; he does not mention the geometrical interpretation of their addition and multiplication. We recall that Argand's *Essai* was originally issued in 1806, and was republished as Argand [1813], and that its ideas were further developed in Argand [1815].

Cauchy begins Section 11 of the memoir by restating, in the geometrical language introduced in Section 9, his results for integrals along two neighbouring paths; he now expresses them as follows. If $f(x + y i)$ is never infinite on or between the paths, thus the integrals along the two paths are equal; if, however, there are points between the paths where $f(x + y i)$ becomes infinite, then the integrals will differ by

$$\pm 2\pi f_1 i \pm 2\pi f_2 i \pm \cdots$$

(Section 4.7 above). If such a point lies on one of the paths, the integral along that path will be infinite or indeterminate: if the principal value of the integral exists and is finite, the corresponding term in the above sum has to be halved.

He recalls also that (Section 4.5 above) the ambiguous sign in a term of the sum will be $-$ or $+$ according as

$$x'\delta y - y'\delta x \gtrless 0;$$

if we suppose that $t_0 < T$, this condition can also be written as

$$dx\,\delta y - dy\,\delta x \gtrless 0.$$

He adds the remark that if the two paths do not cross each other, then this sign will be the same for all the points between them.

Cauchy then discusses how to compare the values of the integral along two paths that are not close neighbours; he suggests that one should imagine a variable curve that can be continuously deformed, initially coinciding with one of the paths and finally with the other one; in this way one can establish that in this case also the difference between the integrals, provided that it is finite, will be a sum of terms of the form $\pm 2\pi f i$ corresponding to points lying between the paths and of the form $\pm \pi f i$ for points lying on one of them[1]; furthermore, this will hold also for paths made up of portions defined by different functions, a situation that he returns to in Section 16 (Section 4.12 below). He introduces the word 'contour' here to describe such paths, and remarks that the result is still valid even if the contour goes outside the rectangle with opposite vertices at (x_0, y_0) and (X, Y), but insists that in this case the integral along it is not to be regarded as an admissible value of

$$\int_{x_0 + y_0 i}^{X + Y i} f(z)\,dz.$$

We note that such a contour could not be monotonic.

One naturally asks why Cauchy should suddenly introduce the use of geometrical language at this point in the memoir; in spite of his previous mistrust of it. One can suggest several possible reasons: it enabled him (i) to talk about a point lying between two paths, (ii) to clarify the way in which the ambiguous signs in the terms $\pm 2\pi f i$ are to be determined, and in particular to deal with cases where the two paths cross each other, (iii) to describe more clearly the way in which a contour can be built up of portions defined by different functions, and (iv) to describe his homotopy argument for deforming one path into another. In all these cases the use of geometrical language provides a more lucid description of the situation. In the long run, his adoption of geometrical modes of expression, as we shall see in Section 5.10 below, was to pay handsome dividends by clarifying and simplifying the treatment of certain situations.

4.9. In Section 10 of the 1825 memoir Cauchy picks out certain particular paths joining (x_0, y_0) and (X, Y) as deserving special attention. If the path is the straight segment joining these points (the diagonal of the rectangle), he calls the integral along it the *mean value* of

$$\int_{x_0 + y_0 i}^{X + Y i} f(z)\,dz.$$

If, on the other hand, the path is formed by taking either the sides $y = y_0$,

---

[1] He ignores the possibility that $f(z)$ may become infinite at a corner where the path abruptly changes direction, a situation he had examined briefly in Part II, Section IV of the 1814 memoir (Section 2.12 above).

$x = X$ or the sides $x = x_0$, $y = Y$ of the rectangle, the corresponding values of the integral are its *extreme values*.

In his Section 15 he compares the mean value with one of the extreme values; he then allows X to become infinite, assuming that $f(x + yi)$ vanishes when x goes to infinity, thus obtaining results such as

$$\int_0^\infty x^{a-1} e^{-x\cos\theta} \cos(x\sin\theta)\,dx = \cos a\theta \cdot \int_0^\infty x^{a-1} e^{-x}\,dx \qquad (a > 0).$$

There is a similar result in the 1814 memoir (Section 2.6 above), originally given by Euler (Section 1.9 above).

In his Section 12 he compares the two extreme values of the integral, and so obtains a slightly rewritten form of his standard equation (cf. Section 3.5 above)

$$\int_{x_0}^X f(x + y_0 i)\,dx + i\int_{y_0}^Y f(X + yi)\,dy = \int_{x_0}^X f(x + Yi)\,dx$$

$$+ i\int_{y_0}^Y f(x_0 + yi)\,dy + \Delta, \qquad (4.6)$$

where

$$\Delta = 2\pi i(f_1 + f_2 + \cdots),$$

the terms f_1, f_2, ... arising from the infinities of $f(z)$ inside the rectangle. He draws attention to the fact that he is no longer requiring $f(z)$ to be real for real z, and he gives a careful account of the precautions that have to be taken if $f(z)$ involves potentially many-valued expressions such as $(u + vi)^\mu$ or $\log(u + vi)$.

In cases when X or Y becomes infinite he imposes his usual inadequate conditions (Section 3.9 above) on the behaviour of $f(x + yi)$ for large x or y. As we shall see shortly (Section 4.11 below) Cauchy confesses in his Section 14 that these conditions are not always strong enough.

The rest of Section 12 contains many evaluations of definite integrals and derivations of identities between definite integrals. Some of his results involve the gamma function, e.g.

$$\int_{-\infty}^\infty \frac{dx}{(r + xi)^a(s - xi)^b} = 2\pi(r + s)^{1-a-b} \cdot \frac{\Gamma(a + b - 1)}{\Gamma(a)\Gamma(b)},$$

where a, b, r, s are all positive. He remarks that additional results can be obtained by differentiating formulae with respect to a parameter, or by a change of variable, e.g. by the substitution $x = \tan p$.

He mentions that in some cases the equation $1/f(z) = 0$ may have an infinity of roots, so that the sum

$$\Delta = 2\pi i(f_1 + f_2 + \cdots)$$

becomes an infinite series; he examines this possibility more deeply in his Section 13 (Section 4.10 below).

4.10. In Section 13 of the 1825 memoir, Cauchy considers how his results can be applied to the summation of infinite series. He begins by remarking that if $f(x + yi)$ vanishes when $x = \pm\infty$ for all y and when $y = \pm\infty$ for all x, then it will follow from equation (4.6) above that $\Delta = 0$ for the whole plane. He recalls that he had obtained an equivalent result for rational functions in an earlier paper;[2] the proof given there is of a purely algebraic character.

He now suggests that if $1/f(z) = 0$ has an infinity of roots, this result can be used to evaluate the sums of infinite series. As an illustration, he takes $f(x)$ to be of one of the forms

$$\varphi(x)\frac{\cos rx}{\sin \pi x}, \quad \varphi(x)\frac{\sin rx}{\sin \pi x},$$

where $0 < r < \pi$ and $\varphi(x)$ is a rational function vanishing at infinity; he thus obtains many well-known formulae, such as

$$\frac{\pi \cot \pi u}{2u} = \frac{1}{2u^2} + \sum_{n=1}^{\infty} \frac{1}{u^2 - n^2},$$

whence he derives the equation

$$\sin \pi u = \pi u(1 - u^2)\left(1 - \frac{u^2}{4}\right)\left(1 - \frac{u^2}{9}\right)\cdots.$$

He also obtains Stirling's asymptotic formula[3] for the factorial function (attributing it to Laplace).

We note that in these considerations Cauchy seems to think it is enough to require the function $f(x + yi)$ to tend to 0 in a general way when x or y is large, even though it may have infinities. He gave a much improved account of this type of argument in [1827d] (Section 5.10 below).

4.11. We have already noted (Sections 4.9 and 4.10 above) that in Sections 12 and 13 of the 1825 memoir Cauchy was still assuming his usual inadequate conditions at infinity, and in Section 3.9 above we drew attention to an inconsistency arising as a result.

---

[2] [1813], p. 500, *Oeuvres* **(2)1**, 210.
[3] In Cauchy's equation (152) on p. 46 (*Oeuvres* **(2)15**, 73), $\Gamma(n)$ should read $\Gamma(n + 1)$; the error persists in Darboux's reprinting, but is silently corrected in the German translation, p. 44.

He pursues this question a little further in his Section 14, but without reaching a satisfactory conclusion. He begins by remarking that in the applications he has made of his standard formula for a rectangle (equation (4.6)) he has assumed that the integrals involved in the formula vanish together with their integrands. By this he seems to mean, for instance, that if $f(X + yi)$ vanishes when X becomes infinite, then the same is true for

$$\int_{y_0}^{Y} f(X + yi)\,dy;$$

in other words, he has assumed that he can take the limit under the integral sign. He now goes on to say that this is true in general, but that exceptional cases do exist; the equations then have to be modified.

In particular, the equation $\Delta = 0$ in his Section 13 (Section 4.10 above) is false if $f(x)$ is a rational function in which the degree of the denominator exceeds that of the numerator by exactly unity; he refers to his discussion of this point in [1823a], which we mentioned in Section 3.9 above.

He now suggests a new device for dealing with the difficulty; this is to apply the equations $\Delta = 0$, not to $f(x)$ itself, but to some such function as

$$\frac{f(x)}{1 - \varepsilon x}, \qquad \frac{f(x)}{1 - \varepsilon^2 x^2}, \qquad \frac{f(x)}{1 + \varepsilon^2 x^2},$$

and then to put $\varepsilon = 0$ afterwards. As an illustration, he looks at the formula

$$\sum_{n=1}^{\infty} (-1)^{n-1} \frac{\sin nr}{u^2 - n^2} = -\frac{\pi}{2} \cdot \frac{\sin ru}{\sin \pi u}, \qquad (4.7)$$

obtained by taking

$$f(x) = \frac{x}{u^2 - x^2} \frac{\sin rx}{\sin \pi x}.$$

The fomula (4.7) is correct when $0 \leqslant r < \pi$, but fails for $r = \pi$, where it leads to the equation $0 = 1$; the correct result can be obtained by replacing $f(x)$ by $f(x)/(1 + \varepsilon^2 x^2)$ and putting $\varepsilon = 0$ afterwards.

This is very much an *ad hoc* device; Cauchy is obviously very puzzled by the fact that if a function vanishes for the value $X = \infty$ of a parameter X, its integral may fail to vanish for the same value of the parameter. However, he was very soon to return to the question (cf. Section 4.14 below).

4.12. In Section 16 of the 1825 memoir Cauchy takes up again the idea he had mentioned in his Sections 9 and 11 (Section 4.8 above) that a path of integration may be made up of portions defined by different functions. He now contemplates the possibility of considering paths joining the points (x_0, y_0) and (X, Y) and lying anywhere in the (x, y)-plane; the conditions imposed at the end of his Section 11 that the whole path must lie in the rectangle with its end-points as opposite vertices seems to have disappeared.

He sets up some heavy machinery to describe the integral along such a path. He takes a sequence of points (x_0, y_0), (x_1, y_1), ..., (x_{n-1}, y_{n-1}), (X, Y) along the path, and supposes that the portion of the path between two successive points is defined by a single function or pair of functions. He describes the situation by introducing a pair of functions

$$\varphi(p, q, r, \ldots), \chi(p, q, r, \ldots) \qquad (p_0 \leqslant p \leqslant P, q_0 \leqslant q \leqslant Q, \ldots)$$

such that the portion of the path joining (x_0, y_0) and (x_1, y_1) is given by

$$x = \varphi(p, q_0, r_0, \ldots), \quad y = \chi(p, q_0, r_0, \ldots) \qquad (p_0 \leqslant p \leqslant P),$$

the portion joining (x_1, y_1) and (x_0, y_2) by

$$x = \varphi(P, q, r_0, \ldots), \quad y = \chi(P, q, r_0, \ldots) \qquad (q_0 \leqslant q \leqslant Q)$$

and so on. He then writes the integral along the path in terms of these parametrisations in the obvious way.

It seems that Cauchy is still uncomfortable with the idea that a function may have different analytic expressions in different intervals; instead he has devised this machinery, which enables him to avoid regarding the whole contour as being defined from beginning to end by a single pair of functions of one real variable.

In fact, Cauchy seems to be aiming mainly at cases where the path falls into two portions, as when it consists of two sides of a rectangle, or when he can take[4]

$$\varphi(p, r) + i\chi(p, r) = r e^{pi} = r(\cos p + i \sin p),$$

thus enabling him to write down the equation

$$\int_{r_0}^{R} e^{Pi} f(r e^{Pi}) \, dr + i \int_{p_0}^{P} r_0 e^{pi} f(r_0 e^{pi}) \, dp = \int_{r_0}^{R} e^{p_0 i} f(r e^{p_0 i}) \, dr$$

$$+ i \int_{p_0}^{P} R e^{pi} f(R e^{pi}) \, dp + \Delta,$$

where

$$\Delta = -2\pi i (f_1 + f_2 + \cdots),$$

---

[4] In the original paper [1825b, p. 53] and in *Oeuvres* **(2)15**, 79, $f(p, r)$ appears instead of $\varphi(p, r) + i\chi(p, r)$; the error persists in Darboux's reprinting, but is silently corrected in the German translation, p. 51.

the terms f_1, f_2, ... arising from infinities of $f(z)$ at values of $z = r\,e^{pi}$ with r between r_0 and R and p between p_0 and P.

Apart from some details of notation, he had obtained this equation earlier[5] (cf. Section 3.7 above) by a different method, using the generalised Cauchy–Riemann equations of the 1814 memoir (Section 2.3 above), which we interpreted in Section 2.4 in terms of a change of coordinates. This complication has now disappeared, everything being now expressed in terms of integrals along paths in the (x, y)-plane.

By specialising the values of r_0, R, p_0 and P he derives formulae similar to those obtained in [1822b] and [1823a]. One such result is

$$\int_0^\pi e^{pi} f(e^{pi})\,\mathrm{d}p = i\int_{-1}^1 f(r)\,\mathrm{d}r + \Delta i, \qquad (4.8)$$

a special case of which was given by Vernier [1824].

In cases where $f(z)$ has no infinities with $|z| < 1$, he obtains a group of formulae of the type

$$\int_0^{\pi/2} \frac{s\sin 2p}{1 - 2s\cos 2p + s^2}\,\frac{f(e^{2pi}) - f(e^{-2pi})}{2i}\,\mathrm{d}p = \frac{\pi}{4}[f(s) - f(0)]$$

$$(0 < s < 1).$$

He mentions that these could also be obtained directly from the theorem of Parseval [1805] (cf. Section 3.7 above), and refers to other similar formulae given by Libri [1822], Poisson [1822] and himself [1822b].

Although Cauchy says nothing explicitly, he appears to have tacitly abandoned his restriction to monotonic paths; the integral on the left-hand side of (4.8) above is along a non-monotonic path (although it could be broken up into two successive monotonic paths).

4.13. The remainder of the 1825 memoir contains little of interest from the point of view of general theory.

In Section 17 Cauchy mentions the possibility of extending his methods to deal with multiple integrals with real or imaginary limits, but does not develop the idea, promising to do so in a later paper; nothing appeared on this in the period up to 1831.

In Section 18 he is concerned with a special integral arising in his work on water waves [1815]. Here he applies the formula

$$\int_0^\infty f(x)\,\mathrm{d}x = i\int_0^\infty f(yi)\,\mathrm{d}y,$$

---

[5] [1823a], p. 575; *Oeuvres* (2)**1**, 338.

valid for a function with no infinities in the first quadrant, to the function

$$f(x) = x \, e^{x^2 i} \, e^{axi};$$

this does not vanish when x becomes infinite, so that even his usual inadequate conditions are not satisfied. He is interested in finding an asymptotic expression for the integral

$$\int_0^\infty a \cos{(a\alpha)} \cos{(\alpha^2)} \, d\alpha,$$

which is easily seen not to be convergent. However, Cauchy is not, in general, as insistent on convergence for infinite integrals as he is for infinite series,[6] being prepared to use some summability device for their evaluation.

The memoir ends with a five-page 'Addition', consisting of a long list of definite integrals; he gives values for many of these, stating that the others can be evaluated by similar methods. He remarks that in some of the formulae parameters initially assumed to be real can be allowed to take suitable imaginary values.

4.14. We now discuss a memoir that was published in two parts, [1825c] and [1826f], in Gergonne's *Annales de Mathématiques*. It is essentially a pendant to the 1825 memoir, to which there are a number of references, and in it Cauchy makes a serious attempt to strengthen his conditions at infinity in order to deal with the problem discussed in Section 14 of the 1825 memoir (Section 4.11 above).

In the first part of this memoir he recalls his standard formula

$$\int_{-\infty}^\infty f(x) \, dx = \Delta = 2\pi i (f_1 + f_2 + \cdots), \tag{4.9}$$

which he had stated under his usual inadequate conditions at infinity; the constants $f_1, f_2 \ldots$ arise from those roots of the equation $1/f(x) = 0$ that have positive imaginary parts. He now proposes to give a direct proof of (4.9), in view of its usefulness for the evaluation of definite integrals.

His main result is as follows.

Theorem–Suppose that $f(x + yi)$ vanishes when $x = \pm\infty$ for all y and when $y = +\infty$ for all x, and is finite and continuous for x between $-\infty$ and $+\infty$ and y between 0 and $+\infty$, and suppose also that $xf(x)$ tends to F when the numerical value of x becomes infinite; then

$$\int_{-\infty}^\infty f(x) \, dx = -\pi F i.$$

[6] Smithies [1986], p. 57.

We note that he is explicitly assuming only that $xf(x) \to F$ as $x \to \pm\infty$ through real values, in fact, what he uses later is that $(x + yi)f(x + yi) \to F$ as $x \to \pm\infty$.

To prove the result, he argues as follows. Since

$$\frac{\partial f(x + yi)}{\partial y} = i\frac{\partial f(x + yi)}{\partial x},$$

we have

$$\int_{-X}^{X} dx \int_0^\infty \frac{\partial f(x + yi)}{\partial y} \, dy = i\int_0^\infty dy \int_{-X}^{X} \frac{\partial f(x + yi)}{\partial x} \, dx;$$

since $f(x + yi)$ vanishes when y becomes infinite, this reduces to

$$\int_{-X}^{X} f(x) \, dx = -i\int_0^\infty [f(X + yi) - f(-X + yi)] \, dy.$$

For large X he uses the approximations

$$f(X + yi) = \frac{F}{X + yi}, \quad f(-X + yi) = \frac{F}{-X + yi},$$

whence we have nearly

$$\int_{-X}^{X} f(x) \, dx = -Fi\int_0^\infty \left(\frac{1}{X + yi} - \frac{1}{-X + yi}\right) dy$$

$$= -\pi F i.$$

The last equation becomes exact when $X = +\infty$, and we have the required result.

He then states two immediate corollaries: (i) if $F = 0$, then

$$\int_{-\infty}^{\infty} f(x) \, dx = 0;$$

(ii) if $f(x) - \varphi(x)$ satisfies the conditions of the theorem, and $xf(x)$ and $x\varphi(x)$ tend to F and Φ, respectively, when x becomes infinite, then

$$\int_{-\infty}^{\infty} f(x) \, dx = \int_{-\infty}^{\infty} \varphi(x) \, dx + \pi i(\Phi - F).$$

As a third corollary, he restates his standard formula

$$\int_{-\infty}^{\infty} f(x) \, dx = 2\pi i(f_1 + f_2 + \cdots),$$

where $f(x + yi)$ is now required to satisfy both his usual conditions at infinity and also

$$xf(x) \to 0 \qquad (x \to \pm\infty).$$

To obtain this result, he constructs a rational function $\varphi(x)$ such that $f(x) - \varphi(x)$ satisfies the conditions of the main theorem.

For instance, if x_1 is the only infinity of $f(x)$ with a positive imaginary part, and

$$(x - x_1)f(x) \rightarrow f_1 \qquad (x - x_1 \rightarrow 0),$$

he takes

$$\varphi(x) = \frac{f_1}{x - x_1}.$$

He discusses this special case in some detail, and concludes that here

$$\int_{-\infty}^{\infty} f(x)\,dx = 2\pi i f_1,$$

since $\Phi = f_1$. If, however, x_1 is real, he obtains

$$\int_{-\infty}^{\infty} f(x)\,dx = \pi i f_1,$$

the integral now being a principal value.

More generally, if x_i is a multiple root of order m, we can write (with a slight modification of Cauchy's notation)

$$(x - x_1)^m f(x) = f_0(x)$$

$$= f_0(x_1) + (x - x_1)f_0'(x_1) + \cdots + \frac{(x - x_1)^{m-1}}{(m-1)!}f_0^{(m-1)}(x_1)$$

$$+ (x - x_1)^m \psi(x),$$

where $\psi(x)$ is finite for $x - x_1$. This equation can be rewritten[7] as

$$\psi(x) = f(x) - \frac{f_0(x_1)}{(x - x_1)^m} - \frac{f_0'(x_1)}{(x - x_1)^{m-1}} - \cdots - \frac{f_0^{(m-1)}(x_1)}{(m-1)!}\frac{1}{x - x_1}.$$

so that, in order to apply the second corollary, we can take

$$\varphi(x) = \frac{f_0(x)}{(x - x_1)^m} + \frac{f_0'(x_1)}{(x - x_1)^{m-1}} + \cdots + \frac{f_0^{(m-1)}(x_1)}{(m-1)!}\frac{1}{x - x_1}.$$

We therefore have

$$\Phi = \lim_{x \to \pm\infty} x\varphi(x) = \frac{f^{(m-1)}(x_1)}{(m-1)!},$$

so that we still have $\Phi = f_1$, as in the case $m = 1$, provided that we now take

$$f_1 = \frac{f_0^{(m-1)}(x_1)}{(m-1)!} = \lim_{x \to x_1} \frac{1}{(m-1)!}\frac{d^{m-1}}{dx^{m-1}}[(x - x_1)^m f(x)].$$

---

[7] In this and the following equation we have corrected some minor misprints, which are present in both the original paper and the reprint in *Oeuvres* (2)2.

Finally, he observes that if $x_1 = a + bi$, where $b > 0$, we have

$$\int_{-\infty}^{\infty} \frac{dx}{(x - x_1)^k} = 0 \qquad (k > 1),$$

whence

$$\int_{-\infty}^{\infty} \varphi(x)\, dx = f_1 \int_{-\infty}^{\infty} \frac{dx}{x - x_1}$$

$$= f_1 \lim_{X \to \infty} \int_{-X}^{X} \frac{dx}{x - a - bi}$$

$$= f_1 \lim_{X \to \infty} \int_{-X}^{X} \frac{(x - a) + bi}{(x - a)^2 + b^2}\, dx = \pi i f_1;$$

it follows at once that

$$\int_{-\infty}^{\infty} f(x)\, dx = \int_{-\infty}^{\infty} \varphi(x)\, dx + \pi i \Phi$$

$$= 2\pi i f_1.$$

We note that if $F \neq 0$, the full result would be

$$\int_{-\infty}^{\infty} f(x)\, dx = 2\pi i f_1 - \pi i F. \qquad (4.10)$$

He also discusses the case $b = 0$, much as in Section 7 of the 1825 memoir (Section 4.6 above).

He concludes with the usual remarks that if $f(x)$ has several infinities in the upper half-plane, then their contributions have to be summed.

We see that Cauchy has strengthened his conditions at infinity sufficiently to cope with the inconsistencies that had emerged in his earlier work (Section 4.11 above). In later work he was to strengthen them still further (cf. Section 5.10 below).

It is amusing to note that, in order to obtain what is essentially a local result (in his later terminology, the residue of a function at a singularity) Cauchy has here made use of the behaviour of the function $\varphi(x)$ at infinity.

4.15. The second part [1826f] of the memoir in Gergonne's *Annales* is devoted to applications of the main result of the first part, in most cases to the evaluation of various types of definite integrals. Only a few points call for comment.

In recalling the main result of the first part near the beginning of the paper, he imposes only his customary conditions at infinity, although he had recognised their inadequacy in the first part of the memoir. It is only in a footnote[8] inserted in the list of special results that he mentions the necessity of modifying the general result in cases where $xf(x) \to F \neq 0$ as $x \to \pm\infty$; the modified result is that given by equation (4.10) above.

There are a number of results applicable to general functions. For instance, in order to evaluate a class of integrals taken over $(-\pi, \pi)$ or $(0, \pi)$ he starts from the equation

$$\int_{-\infty}^{\infty} \varphi\left(\frac{1+xi}{1-xi}\right) \frac{dx}{x^2+1} = \pi\left(\varphi(0) + \frac{H+Ki}{h+ki} + \cdots\right),$$

where $h + ki$ runs over the values of z with $|z| < 1$ at which $\varphi(z)$ becomes infinite, and $H + Ki$, in language that he had already used elsewhere, but which he does not introduce in the present paper, is the residue of $\varphi(z)$ at $z = h + ki$. Making the substitution $x = \tan\frac{1}{2}p$, he obtains the formula

$$\int_{-\pi}^{\pi} \varphi(e^{pi}) \, dp = 2\pi\left(\varphi(0) + \frac{H+Ki}{h+ki} + \cdots\right),$$

from which he easily derives many of the evaluations of definite integrals that he had stated without proof in [1822b] and [1823a]. He adds the remark that many of the parameters then assumed to be real and positive can be given negative or imaginary values.

As one of his examples he gives the formula (equation (130) in his numbering)

$$f(r) = \frac{1}{2\pi} \int_{-\pi}^{\pi} \frac{f(e^{pi}) \, dp}{1 - r e^{-ip}} \qquad (|r| < 1),$$

which is essentially 'Cauchy's formula' for the value of $f(z)$ at an arbitrary point $z = r$ inside the unit circle. It does not appear, however, that he was as yet thinking in terms of integration round a closed curve (cf. Sections 5.8 and 5.10 below). His comment is that this result provides a method for converting a given function of the variable r into a definite integral whose integrand is a fraction with its numerator independent of r and its denominator a linear function of r. He was later (1831) to use this expression to prove the convergence of the Maclaurin series of $f(z)$ up to the nearest singularity (Section 6.9 below).

The list of evaluations of definite integrals in this paper contains nearly 200 items.

4.16. We now try to summarise the progress that Cauchy has made in this group of papers. He defines definite integrals along paths in the complex

---

[8] [1826f], p. 108; *Oeuvres* (2)2, 371.

plane; initially he requires the paths to be monotonic, but later in the principal 1825 memoir [1825b] he appears to weaken this restriction, allowing contours that are made up of a chain of monotonic paths.

He shows, by an argument based on methods derived from the calculus of variations, that if the integrand is 'finite and continuous', a condition which in his mind implies differentiability, then all admissible paths joining a given pair of points give the same value for the integral; he shows this first for closely neighbouring paths, and then extends it to paths that are not near neighbours by a kind of homotopy argument.

In cases where the integrand becomes infinite at points between two paths, he evaluates the difference between the values of the integral along them in terms of what he was later (cf. Chapter 5 below) to call the residues of the integrand at these singularities. He gives two proofs for the case where the singularities are simple poles, and two also for the case of multiple poles; in each case one proof uses an approximation method and the other, which we conjecture to have been suggested by Ostrogradskiĭ, uses the device of removing the singularity by subtracting an appropriate rational function from the integrand.

He introduces a geometrical representation for complex numbers, representing $x + y\mathrm{i}$ by the point of the plane with Cartesian coordinates (x, y), but makes no use of Argand's geometrical interpretation for the addition and multiplication of complex numbers.

When he is applying his results to the evaluation of infinite integrals, he generally imposes his usual inadequate conditions on the behaviour of the integrands at infinity; at a late stage in the 1825 memoir (Section 4.11 above) he recognises that they involve an assumption that is not always valid about taking limits under the integral sign, and leads at times to erroneous results. In the main 1825 memoir he produces an *ad hoc* and unconvincing device for dealing with the situation, by which he is obviously puzzled and embarrassed. In [1825c], however, he finds a more satisfactory way of dealing with the difficulty, by imposing stronger conditions at infinity (Section 4.14 above). Indeed he succeeds in exploiting this new approach, combined with the 'Ostrogradskiĭ trick', to obtain a new proof of his main results on the evaluation of infinite integrals.

An interesting feature of this group of papers is that one receives an impression that Cauchy is, as it were, thinking aloud; he keeps changing his mind as he goes along, as is witnessed by (i) his introduction and partial abandonment of monotonicity for admissible paths, (ii) his provision of alternative proofs for his main results, (iii) his sudden and unheralded introduction of geometrical language, and (iv) his hesitancy about the introduction of stronger conditions on the behaviour of functions at infinity. It is possible that the tentative form of the exposition may have contributed to his decision to publish the 1825 memoir as a separate pamphlet; he may have intended to produce a more definitive account of these results, later, but never got round to it.

5

The calculus of residues

5.1. In 1826 Cauchy began to publish his *Exercices de Mathématiques*, which was essentially a mathematical periodical consisting entirely of papers written by himself; it appeared at approximately monthly intervals until 1830. He used similar series as vehicles for his publications at later stages of his career; the *Nouveaux Exercices de Mathématiques* were published in Prague in 1835–36 and the *Exercices d'Analyse et de Physique Mathématique* appeared at intervals from 1840 to 1853.

Some of the work appearing in the *Exercices* is expository in character, but much of it contains original research. In particular, as we shall see, his main contributions to what he called the calculus of residues are to be found there.

In the present chapter we shall describe Cauchy's definition of the notion of residue and the way in which he used it for the further development of complex function theory. We shall omit some of the applications he made of it between 1826 and 1830, in particular its use to obtain explicit expressions for solutions of differential equations, and his work on operational methods, which is more closely linked to the theory of Fourier transforms.

5.2. The notion of the residue of a function at a point where the function becomes infinite is defined in Cauchy's paper [1826a] in the first number of the *Exercices*. His definition is as follows. If $f(x)$ becomes infinite at $x = x_1$, then one can expand $f(x_1 + \varepsilon)$ in powers of ε, the expansion beginning with a finite number of terms in negative powers of ε; in modern language, x_1 is a *pole* of the function. He offers no proof of this statement. The *residue* of $f(x)$ relative to x_1 is then defined to be the coefficient of ε^{-1} in this expansion. He draws an analogy between this and the *differential coefficient* of $f(x)$ at a point x_1 in whose neighbourhood

113

the function is well behaved, defined as the coefficient of ε in the expansion of $f(x_1 + \varepsilon)$.

He promises applications to numerous topics, including the Lagrange interpolation formula, the decomposition of a rational function into partial fractions, the evaluation of definite integrals, the summation of infinite series, the integration of linear difference and differential equations, the Lagrange series and other similar series, and the solution of algebraic and transcendental equations.

His first step is to show how residues can be evaluated. If x_1 is a simple root of the equation $1/f(x) = 0$, he argues as follows. We can write (modifying Cauchy's notation slightly)

$$(x - x_1)f(x) = f_0(x),$$

where $f_0(x)$ no longer becomes infinite for $x = x_1$; hence

$$f(x) = \frac{f_0(x)}{x - x_1},$$

$$f(x_1 + \varepsilon) = \frac{1}{\varepsilon}f_0(x_1) + f_0'(x_1 + \theta\varepsilon),$$

where $0 < \theta < 1$. It follows that the residue of $f(x)$ relative to $x = x_1$ is $f_0(x_1)$; in other words, it is the 'true value' of $\varepsilon f(x_1 + \varepsilon)$ for $\varepsilon = 0$.

When x_1 is a multiple root of $1/f(x) = 0$ of order m, say, he proceeds in a similar way. If one writes

$$f(x) = \frac{f_0(x)}{(x - x_1)^m},$$

then

$$f(x_1 + \varepsilon) = \frac{1}{\varepsilon^m}f_0(x_1) + \frac{1}{\varepsilon^{m-1}}\frac{f_0'(x_1)}{1} + \cdots + \frac{1}{\varepsilon}\frac{f_0^{(m-1)}(x_1)}{(m-1)!} + \frac{f_0^{(m)}(x_1 + \theta\varepsilon)}{m!};$$

the residue of $f(x)$ relative to $x = x_1$ is therefore

$$\frac{f_0^{(m-1)}(x_1)}{(m-1)!},$$

which is equal to the value of

$$\frac{1}{(m-1)!}\frac{\mathrm{d}^{m-1}[\varepsilon^m f(x_1 + \varepsilon)]}{\mathrm{d}\varepsilon^{m-1}}$$

for $\varepsilon = 0$.

We note that Cauchy's argument here is based on a form of Taylor's theorem with remainder that is, strictly speaking, valid only for real-valued functions of a real variable, although it would not be difficult to modify it so as to make it rigorous.

We note also that the expressions Cauchy gives here for the residue of a function have already appeared in his earlier work, especially in the 1825 memoir (cf. chapter 4 above) in his evaluations of the difference between the integrals along two paths in the complex plane. What is new here is the identification of the coefficient of ε^{-1} in the expansion of $f(x_1 + \varepsilon)$ as an important property of the function in the neighbourhood of a point where it becomes infinite.

5.3. Cauchy next introduces the notion of the *integral residue* of a function $f(x)$; this is defined to be the sum of its residues relative to all its infinities, and he denotes it by

$$\mathscr{E}((f(x))).$$

He attributes this notation to a suggestion by Ostrogradskiĭ. The letter $\mathscr{E}$ is intended to suggest the word 'extraction'.

He then introduces some elaborations of this notation. For instance, if

$$f(x) = \frac{f_0(x)}{F(x)},$$

then the sum of the residues of $f(x)$ over the zeros of $F(x)$ is to be denoted by

$$\mathscr{E} \frac{f_0(x)}{((F(x)))};$$

on the other hand, the sum of its residues over the infinities of $f_0(x)$ will be

$$\mathscr{E} \frac{((f_0(x)))}{F(x)}.$$

In particular, the residue of

$$f(x) = \frac{f_0(x)}{(x - x_1)\varphi(x)}$$

at $x = x_1$ will be

$$\mathscr{E} \frac{f_0(x)}{((x - x_1))\varphi(x)};$$

the residue of a general function $f(x)$ at $x = x_1$ will therefore be

$$\mathscr{E} \frac{(x - x_1)f(x)}{((x - x_1))}.$$

He shows also that if

$$f(x) = \frac{f_0(x)}{F(x)},$$

where $F(x)$ has a simple zero at $x = x_1$, then the residue of $f(x)$ at $x = x_1$ is equal to

$$\frac{f_0(x_1)}{F'(x_1)}.$$

He finally introduces the notation

$$\underset{x_0}{\overset{X}{\mathcal{E}}}\underset{y_0}{\overset{Y}{}}((f(x)))$$

to denote the sum of the residues of $f(x)$ over its infinities with real part between x_0 and X and imaginary part between y_0 and Y. He indicates that such expressions will be (in modern language) additive functions of rectangles, provided that residues relative to infinities on the sides of a rectangle are halved and those relative to infinites at the vertices are divided by 4.

He gives a number of formal properties of the operation of extracting residues. For instance, we have

$$\mathcal{E}((f(x) + g(x))) = \mathcal{E}((f(x))) + \mathcal{E}((g(x)));$$

if the infinities of $f(x, z)$ as a function of x are independent of z, then[1]

$$\mathcal{E}\left(\left(\frac{\partial f(x, z)}{\partial z}\right)\right) = \frac{\partial}{\partial z}\mathcal{E}((f(x, z)));$$

and, under similar conditions, we shall have

$$\int_{z_0}^{z} \mathcal{E}((F(x, z)))\, dz = \mathcal{E}\left(\left(\int_{z_0}^{z} F(x, z)\, dz\right)\right).$$

In the 1825 memoir Cauchy had shown that if $f(x)$ becomes infinite at $x = x_1$, then the infinity can be removed by subtracting a simple rational expression (cf. Sections 4.5 and 4.6 above). He now shows that, in his new notation, the expression to be subtracted is

$$\mathcal{E}\frac{(z - x_1)f(z)}{(x - z)((z - x_1))},$$

where the operator $\mathcal{E}$ is taken with respect to the variable z; if one wishes to remove all the infinites at once, then one has to subtract

$$\mathcal{E}\frac{((f(z)))}{x - z}.$$

As an immediate application of this result, he remarks that if

$$f(x) = \frac{f_0(x)}{F(x)}$$

---

[1] In the original version of [1826a], $df(x, z)/dz$ is misprinted as $df(x, z)/dx$; the error is corrected in an errata sheet bound in with this volume of the *Exercices*, and all the corrections mentioned there are taken into account in *Oeuvres* (2)6: from now on we shall not mention these corrections.

is a quotient of polynomials, the degree of $f_0(x)$ being smaller than that of $F(x)$, then we shall have

$$f(x) = \mathscr{E}\frac{((f(z)))}{x - z}, \tag{5.1}$$

which is precisely the expansion of $f(x)$ in partial fractions. He gives an example showing how this works in practice. In particular, if

$$f(x) = \frac{f_0(x)}{(x - x_1)(x - x_2)\dots(x - x_m)},$$

where $x_1, x_2, \dots, x_m$ are distinct, then it follows that

$$f_0(x) = \frac{(x - x_2)\dots(x - x_m)}{(x_1 - x_2)\dots(x_1 - x_m)}f_0(x_1)$$

$$+ \frac{(x - x_1)(x - x_3)\dots(x - x_m)}{(x_2 - x_1)(x_2 - x_3)\dots(x_2 - x_m)}f_0(x_2)$$

$$+ \cdots + \frac{(x - x_1)\dots(x - x_{m-1})}{(x_m - x_1)\dots(x_m - x_{m-1})}f_0(x_m),$$

which is just Lagrange's interpolation formula.

Finally, he considers the case where $f(x)$ is rational and $xf(x)$ takes the value F when x becomes infinite; equation (5.1) above gives

$$xf(x) = \mathscr{E}\frac{((f(z)))}{1 - (z/x)};$$

from this, by letting x become infinite, he concludes that

$$\mathscr{E}((f(x))) = F.$$

He suggests that this result will hold in other cases; he probably has in mind situations like that mentioned in Section 4.10 above, involving functions that become infinite at a sequence of points tending to infinity. A much improved treatment of this situation was ultimately to appear in [1827d] (cf. Section 5.10 below). In any case, he was not at this time aware that if $f(x)$ is meromorphic and $xf(x)$ has a finite limit when $x \to \infty$, then $f(x)$ must be a rational function.

5.4. The next note to be considered is [1826b]. Cauchy begins by recalling from Lessons 33 and 34 of his *Calcul infinitésimal* (cf. Section 3.6 above) that, if

$$\varphi(x, y) = f(z)\frac{\partial z}{\partial x}, \quad \chi(x, y) = f(z)\frac{\partial z}{\partial y},$$

then (under the tacit assumption that $f'(z)$ exists) one has

$$\frac{\partial \varphi(x, y)}{\partial y} = \frac{\partial \chi(x, y)}{\partial x} = F(x, y),$$

say. If $F(x, y)$ is continuous, then

$$\int_{x_0}^{X} \int_{y_0}^{Y} F(x, y) \, dy \, dx = \int_{y_0}^{Y} \int_{x_0}^{X} F(x, y) \, dx \, dy,$$

whence

$$\int_{x_0}^{X} [\varphi(x, Y) - \varphi(x, y_0)] \, dx = \int_{y_0}^{Y} [\chi(X, y) - \chi(x_0, y)] \, dy. \tag{5.2}$$

If, however, $F(x, y)$ has an infinity (or an indeterminacy) between these limits, then a correction term will in general have to be subtracted from the right-hand side of equation (5.2).

In the present note he is interested in the particular case where $z = x + y\mathrm{i}$ and

$$f(z) = \frac{1}{(z - a - b\mathrm{i})^m},$$

where m is a positive integer, so that

$$\varphi(x, y) = \frac{1}{[(x - a) + (y - b)\mathrm{i}]^m}, \quad \chi(x, y) = \frac{\mathrm{i}}{[(x - a) + (y - b)\mathrm{i}]^m}.$$

He evaluates the correction term Δ for this case by explicitly carrying out the necessary integrations. We shall give only his results.

In the case $m = 1$, he shows that

$$\Delta = 2\pi\mathrm{i}$$

if (a, b) is inside the rectangle and $\Delta = 0$ if it is outside.[2] If (a, b) lies on one side of the rectangle, but is not at a vertex, then $\Delta = \pi\mathrm{i}$, the integral along that side being interpreted as a principal value. If (a, b) is at a vertex, then $\Delta = \frac{1}{2}\pi\mathrm{i}$, and the intergals now have to be interpreted as what one might call a 'biprincipal value'; thus, for the case $a = x_0$, $b = y_0$, the difference between the two integrals would have to be taken as

$$\lim_{\varepsilon \to 0+} \left\{ \int_{y_0+\varepsilon}^{Y} [\chi(X, y) - \chi(x_0, y)] \, dy - \int_{x_0+\varepsilon}^{X} [\varphi(x, Y) - \varphi(x, y_0)] \, dx \right\}.$$

In the case $m > 1$, he finds that $\Delta = 0$ if (a, b) is either inside or outside the rectangle; if it lies on one side, but not at a vertex, then Δ is infinite if m is even, but $\Delta = 0$ if m is odd; if (a, b) is at a vertex, then Δ is infinite unless m is of the form $4k + 1$, in which case $\Delta = 0$ and we have to treat

---

[2] Cauchy is not using geometrical language in the present note, but its use enables us to express his results more concisely.

the difference between the integrals as a biprincipal value in the sense described above.

In his later work, Cauchy gives up taking account of infinities on the boundary in the way he had done hitherto, except for a passing remark in [1826g] (Section 5.8 below).

The note [1826b] is essentially a lemma for [1826c]. The arguments he uses to establish the result bear a close similarity to the second proof in Section 5 of the 1825 memoir and the first proof in Section 6 (Sections 4.5 and 4.6 above). We recall that those proofs did not rest on approximation procedures, as his earlier evaluations of the correction term had done, but were at that stage stated only for closely neighbouring paths; he has now shown that he can obtain the results directly for well-separated paths, although in this case only for paths along the sides of a rectangle.

In the next note [1826c] Cauchy applies the results of [1826b] to express correction terms in the language of residues. He wishes to evaluate

$$\Delta = i \int_{y_0}^{Y} [f(X + yi) - f(x_0 + yi)] \, dy - \int_{x_0}^{X} [f(x + Yi) - f(x + y_0 i)] \, dx.$$

He showed in [1826a] (Section 5.3 above) that if

$$\varpi(x) = f(x) - \mathcal{E} \frac{((f(z)))}{x - z},$$

then $\varpi(x)$ has no infinities, and so contributes nothing to the correction term; we can therefore replace $f(x)$ by the rational function

$$\mathcal{E} \frac{((f(z)))}{x - z}.$$

If now $a + bi$ is a simple zero of $1/f(z)$, it follows at once from [1826b] that its contribution to Δ is $2\pi i$ times the residue of $f(z)$ at $a + bi$ if (a, b) is inside the rectangle and 0 if it is outside, with the usual supplementary remarks if (a, b) is on a side or at a vertex. If all the zeros of $1/f(z)$ are simple, we can conclude at once that

$$\Delta = 2\pi i \, {}_{x_0}^{X} \mathcal{E} \, {}_{y_0}^{Y} ((f(z))).$$

He now turns to the case where $z = a + bi$ is an m-fold root of $1/f(z) = 0$. He begins by remarking that $f(z)/(x - z)$ will have the same residue at $z = a + bi$ as the function

$$\frac{f(z)}{x - z} - \frac{(z - a - bi)^m}{(x - a - bi)^m} \frac{f(z)}{x - z},$$

which can also be written in the form

$$\frac{f(z)}{x - a - bi} + \frac{(z - a - bi)f(z)}{(x - a - bi)^2} + \cdots + \frac{(z - a - bi)^{m-1}f(z)}{(x - a - bi)^m},$$

the factor $(x - z)$ having now disappeared from the denominator. The residue of $f(z)/(x - z)$ at $z = a + bi$ will therefore be

$$\frac{A_1}{x - a - bi} + \frac{A_2}{(x - a - bi)^2} + \cdots + \frac{A_m}{(x - a - bi)^m},$$

where A_k is the residue of $(z - a - bi)^{k-1} f(z)$ $(k = 1, 2, \ldots, m)$. The correction term for

$$\mathscr{E}\frac{((f(z)))}{x - z}$$

will then be

$$A_1 s_1 + A_2 s_2 + \cdots + A_m s_m,$$

where s_k is the correction term for $1/(x - a - bi)^k$ $(k = 1, 2, \ldots, m)$. It now follows from [1826b] that in this case also we shall have

$$\Delta = 2\pi \mathrm{i} \, _{x_0}^{X} \mathscr{E} \, _{y_0}^{Y}((f(z))),$$

the usual adjustments being made for infinities on a side or at a vertex of the rectangle.

5.5. In the last part of [1826c] Cauchy considers some questions concerned with infinite integrals.

For his main result, his assumption is that $f(x + yi)$ vanishes for all y when $x = \pm\infty$ and for all x when $y = +\infty$, and in addition that $(x + yi)f(x + yi)$ has the limit F when y goes to $+\infty$. He argues that if b is large then $f(x + bi)$ is practically equal to $F/(x + bi)$, so that we have approximately

$$\int_{-\infty}^{\infty} f(x + bi)\, dx = F \int_{-\infty}^{\infty} \frac{dx}{x + bi}$$

$$= F \int_{-\infty}^{\infty} \frac{x\, dx}{x^2 + b^2} - F\mathrm{i} \int_{-\infty}^{\infty} \frac{b\, dx}{x^2 + b^2}$$

$$= -\pi F\mathrm{i},$$

the first of the two integrals on the right-hand side being interpreted as a principal value. He concludes that

$$\int_{-\infty}^{\infty} f(x)\, dx = 2\pi \mathrm{i}\{ _{-\infty}^{\infty}\mathscr{E} _{0}^{\infty}((f(z))) - \tfrac{1}{2}F\}. \tag{5.3}$$

Alternatively, he argues that, having established for the case $F = 0$ that

$$\int_{-\infty}^{\infty} f(x)\, dx = 2\pi \mathrm{i} \, _{-\infty}^{\infty}\mathscr{E} _{0}^{\infty}((f(z))),$$

we can reduce the general case to this by replacing $f(x + yi)$ by $f(x + yi) - F/(x + yi)$, and thus obtain the same result.

We note that, although his conditions on the behaviour of $f(x + yi)$ at infinity have become stricter, they still do not cover all possibilities. For example, the function e^{-z^4} would satisfy all his conditions, but his conclusion would give the obviously erroneous result

$$\int_{-\infty}^{\infty} e^{-x^4}\, dx = 0.$$

After giving some examples of the use of these results for the evaluation of definite integrals and, incidentally, correcting some erroneous formulae stated in the 1825 memoir, Cauchy considers some further questions associated with the behaviour of functions at infinity. He now supposes that $f(x + yi)$ vanishes for all y when $x = \pm\infty$ and for all x when $y = \pm\infty$, and that $(x + yi)f(x + yi)$ reduces to F under the same conditions. He then argues that if x_0 and y_0 are large and negative and X and Y are large and positive, then we shall have approximately

$$i\int_{y_0}^{Y} [f(X + yi) - f(x_0 + yi)]\, dy - \int_{x_0}^{X} [f(x + Yi) - f(x + y_0 i)]\, dx$$

$$= i\int_{y_0}^{Y} \left(\frac{F}{X + yi} - \frac{F}{x_0 + yi}\right) dy - \int_{x_0}^{X} \left(\frac{F}{x + Yi} - \frac{F}{x + y_0 i}\right) dx$$

$$= 2\pi i F;$$

hence, if we take $x_0 = y_0 = -\infty$, $X = Y = +\infty$, we shall have

$$2\pi i \mathscr{E}((f(z))) = 2\pi i F,$$

i.e.

$$\mathscr{E}((f(z))) = F.$$

In particular, if $F = 0$ then $\mathscr{E}((f(z))) = 0$.
    He next attempts to show that if

$$\lim_{x \to \infty} (x + yi)f(x + yi) = F_1, \quad \lim_{x \to -\infty} (x + yi)f(x + yi) = F_2,$$

then

$$\mathscr{E}((f(z))) = \tfrac{1}{2}(F_1 + F_2).$$

From this he concludes that if $f(x + yi)$ remains finite when x or y goes to $\pm\infty$, tending to F_1 when $x \to +\infty$ and to F_2 when $x \to -\infty$, where $F_1 + F_2 = 0$, then it will follow by applying his result to $f(z)/(x - z)$, that

$$f(x) = \mathscr{E}\frac{((f(z)))}{x - z}.$$

He then suggests that we can use this result to deduce such formulae as

$$\cot x = \mathscr{E}\,\frac{\cos z}{(x - z)((\sin z))}$$

$$= \frac{1}{x} - 2x\left(\frac{1}{\pi^2 - x^2} + \frac{1}{4\pi^2 - x^2} + \frac{1}{9\pi^2 - x^2} + \cdots\right).$$

In all this he takes no notice of the fact that his conditions at infinity are not satisfied, because the function $\cot x$ has infinities for arbitrarily large values of x.

He was to return to this topic in [1827d] (Section 5.10 below), where he recognises that his equation

$$\mathscr{E}((f(z))) = F$$

requires much more careful interpretation than he gives it here; by this time he had also discovered the advantages of working with circles rather than rectangles.

5.6. We next consider the notes [1826d] and [1826e], in which Cauchy examines some formal properties of residues; these were to be used later as lemmas when he began to work in terms of polar coordinates (Section 5.9 below).

In [1826d] he recalls from [1826c] his result that if $zf(z) \to F$ when the real and imaginary parts of z become infinite, then

$$\mathscr{E}((f(z))) = F.$$

He now makes the substitution $z = 1/u$, and claims that his assumption implies that F will be the value of $(1/u)f(1/u)$ at $u = 0$; equivalently, F will be the residue of $(1/u^2)f(1/u)$ at $u = 0$, so that

$$\mathscr{E}((f(z))) = \mathscr{E}\,\frac{f(1/u)}{((u^2))}.$$

He proposes to extend this result to more general functions; he assumes that $f(1/u)/u^2$ has a finite residue at $u = 0$. Supposing that $f(1/u)$ has an infinity of order m at $u = 0$, he argues as follows. We can write

$$\frac{1}{u^2}f\left(\frac{1}{u}\right) = \mathscr{E}\,\frac{f(1/s)}{((s^2))(u - s)} + U(u),$$

where U is finite at $u = 0$. Expanding $1/(u - s)$ in powers of s, he deduces that

$$\frac{f(1/u)}{u^2} = \frac{1}{u}\mathscr{E}\,\frac{f(1/s)}{((s^2))} + \frac{1}{u^2}\mathscr{E}\,\frac{sf(1/s)}{((s^2))} + \cdots + \frac{1}{u^{m-2}}\mathscr{E}\,\frac{s^{m+1}f(1/s)}{((s^2))} + U(u),$$

since the sign $\mathscr{E}$ kills all the higher terms of the series. He rewrites this equation as

$$f(z) = \frac{1}{z}\mathscr{E}\frac{f(1/s)}{((s^2))} + \mathscr{E}\frac{f(1/s)}{((s))} + \cdots + z^m\mathscr{E}\frac{s^m f(1/s)}{((s))} + \varpi(z),$$

where $z\varpi(z)$ vanishes at infinity. Taking integral residues with respect to z in this equation, he concludes that

$$\mathscr{E}((f(z))) = \mathscr{E}\frac{f(1/s)}{((s^2))}, \tag{5.4}$$

since all the other terms vanish; this is the required result.

He gives an alternative proof, which involves a rather messy argument that would be difficult to justify.

Replacing $f(z)$ in (5.4) by $f(z)/(z-x)$, Cauchy deduces the formula

$$f(x) = \mathscr{E}\frac{((f(z)))}{x-z} + \mathscr{E}\frac{f(1/z)}{((z))(1-zx)}, \tag{5.5}$$

valid now, he says, for functions that become infinite with the real and imaginary parts of z. He remarks that if $f(z)$ is a polynomial, then (5.5) reduces to

$$f(x) = \mathscr{E}\frac{f(1/z)}{((z))(1-zx)};$$

more generally, if $f(x)$ is a quotient of polynomials, say

$$f(x) = f_0(x)/F_0(x),$$

where

$$f_0(x) = q(x)F_0(x) + r(x),$$

$r(x)$ being of lower degree than $F_0(x)$, then the second term on the right-hand side of (5.5) is equal to $q(x)$, and the first term gives the partial-fraction expansion of $r(x)/F_0(x)$. He gives an illustrative example.

He applies (5.5) to the function $f(z) = \cot(1/z)$, regardless of the fact that the singularity at $z = 0$ is not isolated, and so obtains the familiar expansion

$$\cot\frac{1}{x} = x - 2x\left(\frac{1}{\pi^2 x^2 - 1} + \frac{1}{4\pi^2 x^2 - 1} + \frac{1}{9\pi^2 x^2 - 1} + \cdots\right).$$

He tries to do the same for the function

$$\frac{\sin(a/(x-a))}{\sin(a^2/(x^2 - a^2))},$$

and obtains a result that he later [1827e] recognises to be erroneous.

We note that in equation (5.4) above Cauchy is comparing the behaviour of $f(z)$ for large z with the behaviour of $f(1/s)$ for small s. This move may have contributed to Cauchy's realisation that there would be advantages in working with circles rather than rectangles; this trend becomes explicit, as we shall see, in [1826g] and [1827d] (Sections 5.8 and 5.10 below).

5.7. In [1826e] Cauchy examines the residues of the derivative $f'(z)$ of a function $f(z)$; he then goes on to discuss how residues are transformed when there is a change in the independent variable.

He begins by remarking that if z_1 is an infinity of $f(z)$ and $f_0(z) = (z - z_1)^m f(z)$ vanishes when $z = z_1$ (so that z_1 is an infinity of order less than m), then we have

$$f'(z) = \frac{f_0'(z)}{(z - z_1)^m} - \frac{m f_0(z)}{(z - z_1)^{m+1}};$$

the residue of $f'(z)$ at $z = z_1$ will therefore be

$$\mathscr{E}\,\frac{(z - z_1)f'(z)}{((z - z_1))} = \mathscr{E}\,\frac{f_0'(z)}{(((z - z_1)^m))} - m\,\mathscr{E}\,\frac{f_0(z)}{(((z - z_1)^{m+1}))}$$

$$= \frac{f_0^{(m)}(z_1)}{(m - 1)!} - \frac{m f_0^{(m)}(z_1)}{m!}$$

$$= 0.$$

He gives an alternative proof by expanding $f(z)$ in powers of $(z - z_1)$, pointing out that the differentiated series will contain no term in $1/(z - z_1)$.

From this he concludes that the integral residue of a derivative must vanish. In particular, we have

$$\mathscr{E}((\varphi(z)\chi'(z))) = -\mathscr{E}((\varphi'(z)\chi(z))), \qquad (5.6)$$

with a similar formula for the integral residue over an arbitrary rectangle. He draws an analogy between (5.6) and the familiar formula for integration by parts.

In the remainder of [1826e] he examines what happens to residues under a change of variable. His main result is that if one makes the substitution $z = \psi(t)$, and if z_1 is an infinity of $f(z)$ and $\psi(t_1) = z_1$, then

$$\mathscr{E}\,\frac{(z - z_1)f(z)}{((z - z_1))} = \mathscr{E}\,\frac{(t - t_1)f[\psi(t)]\psi'(t)}{((t - t_1))}, \qquad (5.7)$$

provided that $\psi'(t_1)$ is neither 0 nor infinity; in other words, the residue of $f(z)$ at $z = z_1$ is equal to the residue of $f[\psi(t)]\psi'(t)$ at $t = t_1$.

He proves the result first for the case where z_1 is an infinity of order 1 and then separately when it is of order $m > 1$; we sketch his proof for the latter case. He writes

$$f_0(z) = (z - z_1)^m f(z)$$

and shows that $f(z) = f_0(z)/(z - z_1)^m$ can be expressed in the form

$$f(z) = F'(z) + \frac{1}{(m-1)!} \frac{f_0^{(m-1)}(z_1)}{z - z_1},$$

where $F'(z)$ is a derivative and consequently has residue 0; we therefore have

$$f[\psi(t)]\psi'(t) = F'[\psi(t)]\psi'(t) + \frac{f_0^{(m-1)}(z_1)}{(m-1)!} \cdot \frac{\psi'(t)}{\psi(t) - \psi(t_1)}. \tag{5.8}$$

He then argues that, since $\psi'(t)$ is neither zero nor infinite when $t = t_1$, the same is true of

$$\frac{\psi(t) - \psi(t_1)}{t - t_1},$$

so that we can write

$$\varpi(t) = \log \frac{\psi(t) - \psi(t_1)}{t - t_1},$$

where $\varpi(t)$ is well behaved near $t = t_1$; we then have

$$\frac{\psi'(t)}{\psi(t) - \psi(t_1)} = \varpi'(t) + \frac{1}{t - t_1}.$$

Taking residues at $t = t_1$ in (5.8) and observing that $F[\psi(t)]\psi'(t)$ and $\varpi'(t)$ are derivatives, he concludes that the residue of $f[\psi(t)]\psi'(t)$ at $t = t_1$ is equal to

$$\frac{f_0^{(m-1)}(z_1)}{(m-1)!},$$

which is precisely the residue of $f(z)$ at $z = z_1$.

Cauchy also considers what happens[3] if $\psi'(t)$ becomes zero or infinite when $t = t_1$; he supposes that

$$\frac{\psi(t) - \psi(t_1)}{(t - t_1)^\mu}$$

has a non-zero finite value at $t = t_1$. Writing $\varpi(t)$ for the logarithm of this expression, and arguing much as before, he obtains the formula

$$\mathcal{E} \frac{(t - t_1)f[\psi(t)]\psi'(t)}{((t - t_1))} = \mu \mathcal{E} \frac{(z - z_1)f(z)}{((z - z_1))}.$$

---

[3] $\psi'(t)$ is misprinted as $\psi(t)$ in both [1826e], p. 175 and in *Oeuvres* (2)6, 217.

We note that this argument works only if μ is an integer; otherwise one runs into trouble with many-valuedness.

He briefly considers the application of equation (5.7) to integral residues, remarking that if the equation $z = \psi(t)$ has a unique solution t for each z, then

$$\mathscr{E}((f(z))) = \mathscr{E}((f[\psi(t)]\psi'(t)));$$

on the other hand, if $z = \psi(t)$ has m solutions for each z, then

$$\mathscr{E}((f(z))) = m\mathscr{E}((f[\psi(t)]\psi'(t))).$$

He ends [1826e] with an illustrative example, taking

$$f(z) = \frac{1}{1+z^2} f_0\left(\frac{1+zi}{1-zi}\right);$$

he is therefore interested in the effect of the substitution

$$t = \frac{1+zi}{1-zi}, \quad z = \frac{i(1-t)}{1+t}.$$

He concludes that the residue of $f(z)$ at $z = z_1$ is $1/2i$ times that of $f_0(t)/t$ at $t = t_1$. He was to quote this result in [1826g] (Section 5.8 below), where he is concerned with expressing many of his results in terms of polar coordinates.

5.8. We now consider the paper [1826g], in which Cauchy introduces a more general notation for sums of residues; his main aim in so doing is to be able to express his results in terms of polar coordinates.

Supposing now that $z = \varphi(x, y) + i\chi(x, y)$ rather than $z = x + yi$, he introduces the notation

$$\underset{x=x_0}{\overset{x=X}{}}\mathscr{E}\,\underset{y=y_0}{\overset{y=Y}{}}((f(z))) \qquad [z = \varphi(x, y) + i\chi(x, y)]$$

to denote the sum of the residues of $f(z)$ over the roots of $1/f(z) = 0$ for which x lies between x_0 and X and y between y_0 and Y. He suggests some further generalisations, for instance,

$$\underset{x=x_0}{\overset{x=X}{}}\mathscr{E}\,\underset{y=f_0(x)}{\overset{y=F(x)}{}}((f(z))) \qquad [z = \varphi(x, y) + i\chi(x, y)]$$

relates to the infinities of $f(z)$ for which x lies between x_0 and X and y between $f_0(x)$ and $F(x)$. He introduces some geometrical language at this point in order to clarify the situation.

He remarks that the results of [1826e] (Section 5.7 above) about the effect of a substitution $z = x + yi = \psi(t)$ on the residues of a function $f(z)$ can, in the straightforward case where each value of z corresponds to a unique value of t, be written in the form

$$\underset{x_0}{\overset{X}{}}\mathscr{E}\,\underset{y_0}{\overset{Y}{}}((f(z))) = \underset{x=x_0}{\overset{x=X}{}}\mathscr{E}\,\underset{y=y_0}{\overset{y=Y}{}}((f[\psi(t)]\psi'(t))) \qquad [x + yi = \psi(t)]. \qquad (5.9)$$

In spite of this very general set-up, it is clear that Cauchy is primarily interested in the case

$$z = r(\cos p + \mathrm{i} \sin p),$$

so that (r, p) can be interpreted as polar coordinates. For this case he introduces the special notation

$$\overset{(R)}{\underset{(r_0)}{}} \mathscr{E} \overset{(P)}{\underset{(p_0)}{}}((f(z)))$$

to denote the sum of the residues over the infinities of $f(z)$ for which r lies between r_0 and R and p between p_0 and P; it differs from the notation he had used for rectangular coordinates only by the enclosure of the limits r_0, R, p_0, P in parentheses.

To get his results for the case of polar coordinates, he supposes that $\psi(t)$ is a solution of the equation $t = \mathrm{e}^z$ (in other words, a branch of the logarithmic function), where $z = x + y\mathrm{i}$, and writes (apart from a minor change of notation)

$$f(z) = \mathrm{e}^z f_0(\mathrm{e}^z).$$

Equation (5.9) then gives

$$\overset{X}{\underset{x_0}{}} \mathscr{E} \overset{Y}{\underset{y_0}{}}((\mathrm{e}^z f_0(\mathrm{e}^z))) = \overset{x=X}{\underset{x=x_0}{}} \mathscr{E} \overset{y=Y}{\underset{y=y_0}{}}((f_0(t))) \qquad [t = \mathrm{e}^{X+y\mathrm{i}}].$$

He then writes $t = r(\cos p + \mathrm{i} \sin p)$, $\mathrm{e}^x = r$, $y = p$, $\mathrm{e}^{X_0} = r_0$, $y_0 = p_0$, and so on, thus obtaining

$$\overset{X}{\underset{x_0}{}} \mathscr{E} \overset{Y}{\underset{y_0}{}}((\mathrm{e}^z f_0(\mathrm{e}^z))) = \overset{(R)}{\underset{(r_0)}{}} \mathscr{E} \overset{(P)}{\underset{(p_0)}{}}((f_0(t))).$$

Under this transformation, his standard formula

$$\mathrm{i} \int_{y_0}^{Y} [f(X + y\mathrm{i})) - f(x_0 + y\mathrm{i})]\,\mathrm{d}y$$

$$- \int_{x_0}^{X} [f(x + y\mathrm{i}) - f(x + y_0\mathrm{i})]\,\mathrm{d}x = 2\pi\mathrm{i} \overset{X}{\underset{x_0}{}} \mathscr{E} \overset{Y}{\underset{y_0}{}}((f(z)))$$

becomes

$$\mathrm{i} \int_{p_0}^{P} [Rf(R\,\mathrm{e}^{p\mathrm{i}}) - r_0 f(r_0\,\mathrm{e}^{p\mathrm{i}})]\,\mathrm{e}^{p\mathrm{i}}\,\mathrm{d}p$$

$$- \int_{r_0}^{R} [\mathrm{e}^{P\mathrm{i}} f(r\,\mathrm{e}^{P\mathrm{i}}) - \mathrm{e}^{p_0\mathrm{i}} f(r\,\mathrm{e}^{p_0\mathrm{i}})]\,\mathrm{d}r = 2\pi\mathrm{i} \overset{(R)}{\underset{(r_0)}{}} \mathscr{E} \overset{(P)}{\underset{(p_0)}{}}((f(t))). \qquad (5.10)$$

From this equation Cauchy deduces several special cases, e.g.

$$\int_{-\pi}^{\pi} [Rf(R\,\mathrm{e}^{p\mathrm{i}}) - r_0 f(r_0\,\mathrm{e}^{p\mathrm{i}})]\,\mathrm{e}^{p\mathrm{i}}\,\mathrm{d}p = 2\pi \overset{(R)}{\underset{(r_0)}{}} \mathscr{E} \overset{(\pi)}{\underset{(-\pi)}{}}((f(t))).$$

In particular, with $r_0 = 0$, $R = 1$, this reduces to

$$\int_{-\pi}^{\pi} \mathrm{e}^{p\mathrm{i}} f(\mathrm{e}^{p\mathrm{i}})\,\mathrm{d}p = 2\pi \overset{(1)}{\underset{(0)}{}} \mathscr{E} \overset{(\pi)}{\underset{(-\pi)}{}}((f(t))), \qquad (5.11)$$

with the proviso that $tf(t)$ vanishes when $t = 0$; the need for this arises from the fact that he excludes the residue at $t = 0$ from the expresion on the right-hand side of (5.11) because, as he puts it, this point corresponds to $z = -\infty$.

He then remarks that a number of conventions are required for the correct interpretation of equation (5.10). For instance, residues arising from infinities of $f(t)$ on the boundary must be halved (simplicity of the roots being tacitly assumed). This seems to be the last paper in which he mentions this; in later publications he ignores this possibility altogether. Also, if $P - p_0 > 2\pi$, and there are m values of p in the range (p_0, P) corresponding to an infinity of $f(z)$ then the relevant residue has to be multiplied by m.

He next gives an alternative derivation of equation (5.11). He recalls from [1826c] the formula (equation (5.3) above)

$$\int_{-\infty}^{\infty} f(x)\,dx = 2\pi i \,_{-\infty}^{\infty}\mathscr{E}_0^{\infty}((f(z))), \qquad (5.12)$$

valid if $zf(z)$ is small at infinity, so that $F = 0$, and writes (apart from a slight change of notation)

$$f(z) = \frac{1}{1 + z^2} f_0\left(\frac{1 + zi}{1 - zi}\right);$$

under the substitution

$$t = r(\cos p + i \sin p) = \frac{1 + zi}{1 - zi}$$

equation (5.12) becomes

$$\int_{-\infty}^{\infty} f_0\left(\frac{1 + zi}{1 - zi}\right) \frac{dz}{1 + z^2} = \pi \,_{(0)}^{(1)}\mathscr{E}_{(-\pi)}^{(\pi)}\left(\left(\frac{f_0(t)}{t}\right)\right).$$

For real values of z, we have $z = \tan\frac{1}{2}p$; the last equation then becomes

$$\int_{-\pi}^{\pi} f_0(e^{pi})\,dp = 2\pi \,_{(0)}^{(1)}\mathscr{E}_{(-\pi)}^{(\pi)}\left(\left(\frac{f_0(t)}{t}\right)\right). \qquad (5.13)$$

Replacing $f_0(t)$ by $tf_0(t)$, we arrive at equation (5.11) again. The result is still subject to the proviso that the residue of $f_0(t)/t$ at $t = 0$ should vanish. He adds that if the residue has a non-zero value, we have to insert an extra term; so that (5.13) becomes

$$\int_{-\pi}^{\pi} f_0(e^{pi})\,dp = 2\pi\left\{\mathscr{E}\frac{f_0(t)}{((t))} + \,_{(0)}^{(1)}\mathscr{E}_{(-\pi)}^{(\pi)}\left(\left(\frac{f_0(t)}{t}\right)\right)\right\}.$$

If, in particular, $f_0(0)$ is finite, we can replace $\mathscr{E}f_0(t)/((t))$ by $f_0(0)$.

Cauchy's convention that the residue at $t = 0$ should not be included in the sum

$$(1) \underset{(0)}{\mathscr{E}} \underset{(-\pi)}{(\pi)} \left(\left(\frac{f_0(t)}{t} \right) \right)$$

is clearly becoming something of a nuisance. As we shall see (Section 5.10 below) he was to abandon this convention in [1827d]. In the present paper he seems to be treating the point $t = 0$ almost as if it were a boundary point.

We note that he has made no use here of the general results of the 1825 memoir, and he makes but little use of the geometrical language that he introduced there. In the remarks mentioned after equation (5.11) there is a first hint of the notion of winding number; ideas pointing in this direction were to be considerably developed in [1831e] (Section 6.18 below).

The remainder of [1826g] is devoted to a long list of special results. He repeats many results given in [1822b], [1823a] and the 1825 memoir, possibly because he regarded his new treatment of this class of integrals as being more satisfactory than his earlier ones. He mentions that some of the results listed were already known, having been given by Frullani [1818], Poisson [1822] and Libri [1825]; he also remarks, as he had done in [1823a] (Section 3.7 above) that many of them can be derived from Parseval's theorem.

We give a couple of examples of the more interesting special results. If $f(z)$ and $F(z)$ are polynomials, then

$$\int_0^{\pi/2} \frac{f(\cos 2p)}{F(\cos 2p)} \cos^a p \cos ap \, dp$$

$$= \frac{\pi}{2^{a+1}} \left\{ \frac{f(0)}{F(0)} + (1) \underset{(0)}{\mathscr{E}} \underset{-\pi}{(\pi)} \frac{(1+t)^a f(\tfrac{1}{2}(t+t^{-1}))}{t((F(\tfrac{1}{2}(t+t^{-1}))))} \right\} \qquad (a > 0),$$

provided that $f(\tfrac{1}{2}(t+t^{-1}))/F(\tfrac{1}{2}(t+t^{-1}))$ has a finite value at $t = 0$; in particular, we have

$$\int_0^{\pi/2} \cos^{a-1} p \, \frac{\sin ap}{\sin p} \, dp = \frac{\pi}{2},$$

which is independent of a.

Secondly, if n is a positive integer and $2n\pi < s < 2(n+1)\pi$, then

$$\int_{-\pi}^{\pi} \frac{e^{pi} f(e^{-pi}) \, dp}{e^{is(\cos p + i \sin p)} - 1} = \frac{2\pi}{is} \left\{ f(0) + f\left(\frac{2\pi i}{s} \right) + \cdots + f\left(\frac{2n\pi i}{s} \right) \right.$$

$$\left. + f\left(-\frac{2\pi i}{s} \right) + \cdots + f\left(-\frac{2n\pi i}{s} \right) \right\}.$$

From this formula Cauchy deduces, for instance, an expression for

$$\tfrac{1}{2} + e^a + e^{4a} + e^{9a} + \cdots + e^{n^2 a}$$

as a definite integral.

At the end of [1826g] Cauchy mentions further formulae related to some given by Poisson [1823b, p. 494].

5.9. We next consider [1827a], in which Cauchy is interested in finding expressions for sums of the form

$$\varphi(z_1) + \varphi(z_2) + \cdots + \varphi(z_m), \tag{5.14}$$

where $\varphi(z)$ is a well-behaved function and $z_1, z_2, \ldots, z_m$ are the zeros of a function $F(z)$ inside a rectangle[4] or a circle.

He starts from the formula

$$\mathop{X}_{x_0} \mathscr{E} \mathop{Y}_{y_0} \frac{f(z)}{((F(z)))} = \frac{f(z_1)}{F'(z_1)} + \frac{f(z_2)}{F'(z_2)} + \cdots + \frac{f(z_m)}{F'(z_m)},$$

where $z_1, \ldots, z_m$ are the zeros (assumed for the moment to be simple) of $F(z) = F(x + yi)$ with real parts between x_0 and X and imaginary parts between y_0 and Y. To obtain a formula for the expression (5.14) it is obviously desirable to choose the function $f(z)$ so that

$$\frac{f(z_r)}{F'(z_r)} = \varphi(z_r) \qquad (r = 1, 2, \ldots, m).$$

One could achieve this result by taking

$$f(z) = \varphi(z)F'(z)$$

or, more generally,

$$f(z) = \varphi(z)F'(z) - \psi(z)F(z),$$

where $\psi(z)$ is a function that is well behaved at $z_1, \ldots, z_m$. We should then have

$$\varphi(z_1) + \varphi(z_2) + \cdots + \varphi(z_m) = \mathscr{E} \frac{\varphi(z)F'(z) - \psi(z)F(z)}{((F(z)))} \tag{5.15}$$

$$= \mathscr{E} \frac{\varphi(z)F'(z)}{((F(z)))}. \tag{5.16}$$

He goes on to show that (5.15) and (5.16) continue to hold, when interpreted in the usual way, even if some of the zeros $z_1, z_2, \ldots, z_m$ are multiple.

Cauchy then transforms these results in a number of ways. In particular, he uses the results of [1826d] (Section 5.6 above) to show that if

---

[4] No geometrical language appears in this paper.

$$\frac{\varphi(1/z)F'(1/z)}{z^2 F(1/z)}$$

has a finite residue at $z = 0$ and $F(z)$ has only the zeros $z_1, z_2, \ldots, z_m$, then

$$\varphi(z_1) + \varphi(z_2) + \cdots + \varphi(z_m) = \mathcal{E}\frac{\varphi(1/z)F'(1/z)}{((z^2))F(1/z)} - \mathcal{E}\frac{((\varphi(z)F'(z)))}{F(z)};$$

if $\varphi(z)F'(z)$ is well behaved for all z, this reduces to

$$\varphi(z_1) + \varphi(z_2) + \cdots + \varphi(z_m) = \mathcal{E}\frac{\varphi(1/z)F'(1/z)}{((z^2))F(1/z)}.$$

There is a similar formula involving a term in $\psi(z)$, derived from equation (5.15) above.

He examines the special case $\varphi(z) = z^n$, with the aim of finding expressions for

$$s_n = z_1^n + z_2^n + \cdots + z_m^n.$$

Assuming that (i) $F'(z)$ (or $F'(z) - z^{-n}\psi(z)F(z)$) is well behaved for all finite z, (ii)

$$\frac{F'(1/z)}{z^{n+2}F(1/z)} \left(\text{or } \frac{F'(1/z)}{z^{n+2}F(1/z)} - \frac{\psi(z)}{z^2} \right)$$

has a finite residue at $z = 0$, and (iii)

$$\frac{F'(1/z)}{zF(1/z)}$$

takes a finite value a at $z = 0$, he concludes that

$$s_n = -\mathcal{E}\frac{\dfrac{d}{dz}\log\left[z^a F(1/z)\right]}{((z^n))} \qquad (n = 1, 2, \ldots).$$

He also obtains the formula

$$s_{-n} = -\mathcal{E}\frac{\dfrac{d}{dz}\log F(z)}{((z^n))} \qquad (n = 1, 2, \ldots),$$

provided that $F'(z)/z^n F(z)$ vanishes when the real and imaginary parts of z become infinite. He uses these results to obtain some (already known) formulae for s_n when $F(z)$ is a polynomial, in terms of its coefficients.

His next move is bolder. He applies the last result to

$$F(z) = \frac{\sin \pi z}{z}$$

(which does not strictly satisfy his conditions at infinity) with the aim of evaluating

$$1 + \frac{1}{2^n} + \frac{1}{3^n} + \cdots$$

for even values of n; his result is that the sum of the series is equal to the coefficient of z^n in the expansion of

$$-\tfrac{1}{2}n\pi^n \log\left(\frac{\sin z}{z}\right)$$

in ascending powers of z. He deduces a formula for the Bernoulli numbers which, he says, had already been given by Libri.[5]

He then applies a similar technique to the sums of negative powers of the roots of the equation $\tan z = z$. He suggests that if these roots are $\alpha, \beta, \gamma, \ldots$, they could be evaluated by noting that

$$\alpha^2 = \lim_{n \to \infty} \frac{S_{-2n}}{S_{-2n-2}},$$

$$\beta^2 = \lim_{n \to \infty} \frac{S_{-2n} - \alpha^{-2n}}{S_{-2n-2} - \alpha^{-2n-2}},$$

and so on. He also considers the roots of the equation $\tan z = az$ for various ranges of the parameter a.

In a final section Cauchy turns to questions involving polar coordinates; in this case he converts expressions in terms of residues into definite integrals by using results from [1826g] (Section 5.8 above). His main result is that, if $z_1, z_2, \ldots, z_m$ are the zeros of $F(z)$ with modulus less than unity and $\varphi(z)F'(z)$ is well behaved for $|z| < 1$, then

$$\varphi(z_1) + \varphi(z_2) + \cdots + \varphi(z_m) = \frac{1}{2\pi}\int_0^{2\pi} e^{pi} \frac{\varphi(e^{pi})F'(e^{pi})}{F(e^{pi})}\,dp.$$

It appears from the argument he uses to derive this result that he is still treating $z = 0$ as an exceptional point, as he did in [1826g].

Formulae of this type were to play an important role when Cauchy came to investigate the properties of the Lagrange series and other related series (Chapter 6 below).

5.10. We now come to the important paper [1827d], in which Cauchy makes a really substantial use of geometrical language for the first time since he had introduced it in the 1825 memoir, and in which he begins to work with integrals round closed curves; for the moment these are always circles, which now replace the rectangles on which all his earlier work was based.

---

[5] This paper must be Libri [1825], which was not published until 1831.

He begins by recalling the result stated in [1826c] (Section 5.5 above) that if $f(z)$ vanishes and $zf(z)$ reduces to the constant F when z becomes infinite through real or imaginary values, then

$$\mathscr{E}((f(z))) = F. \tag{5.17}$$

He remarks that this proposition requires much more precise interpretation than it has so far been given.

He first points out that if the equation

$$1/f(z) = 0 \tag{5.18}$$

has an infinity of roots, then the expression

$$\mathscr{E}((f(z)))$$

becomes an infinite series, whose sum may depend on the order in which the terms are taken. To illustrate this remark, he considers the function

$$f(z) = \frac{\pi \cos \pi z}{z \sin \pi z},$$

whose residues are ± 1, $\pm\frac{1}{2}$, $\pm\frac{1}{3}$, ..., $\pm\frac{1}{n}$, ...; he points out that an infinite series comprising these terms may converge to a positive or negative limit or may diverge to $\pm\infty$ (he does not mention the possibility that the partial sums may oscillate), so that the notation

$$\mathscr{E}\left(\left(\frac{\pi \cos \pi z}{z \sin \pi z}\right)\right)$$

does not have a well-defined meaning.

It appears that we have here the first mention in the published literature of this property of conditionally convergent series. Pringsheim [1898, p. 92] states that it first appeared in Cauchy's *Résumés Analytiques*[6] in 1833, but in the passage described above, dating from 1827, the question is discussed in greater depth than in the *Résumés*.

He next suggests that, since every complex number can be represented in the form $z = r e^{pi}$, where r is positive and p is a real number between $-\pi$ and $+\pi$, we should regard (r, p) as being polar coordinates in the plane. If now we are given a family of closed curves, possibly varying in form, whose points recede more and more from the origin, we could consider the sum of the residues of $f(z)$ at the solutions of (5.18) inside each curve in turn, and take the limit of these sums as the curves move towards infinity. In general, however, as he points out, this procedure would still fail to ascribe a definite value to the integral residue

$$\mathscr{E}((f(z))).$$

---

[6] Cauchy [1833b], p. 57; *Oeuvres* **(2)10**, 69–70.

To overcome this difficulty, he suggests that one should restrict one's attention to a family of circles with centre at the origin and with radii tending to infinity; in the notation introduced in [1826g] (Section 5.8 above), this means that we should consider the limit, as R becomes infinite, of the expression

$$\overset{(R)}{\underset{(0)}{}}\mathscr{E}\overset{(\pi)}{\underset{(-\pi)}{}}((f(z))).$$

He proposes to call this limit, if it exists, the *principal value* of $\mathscr{E}((f(z)))$. For example, the principal value of

$$\mathscr{E}\left(\left(\frac{\pi \cos \pi z}{z \sin \pi z}\right)\right)$$

will then be 0.

Cauchy now recalls from [1826g] the equation

$$\int_{-\pi}^{\pi} [Rf(R e^{pi}) - r_0 f(r_0 e^{pi})] e^{pi} \, dp = 2\pi \overset{(R)}{\underset{(r_0)}{}}\mathscr{E}\overset{(\pi)}{\underset{(-\pi)}{}}((f(z))).$$

When $r_0 = 0$, this reduces to

$$\int_{-\pi}^{\pi} Rf(R e^{pi}) e^{pi} \, dp = 2\pi \overset{(R)}{\underset{(0)}{}}\mathscr{E}\overset{(\pi)}{\underset{(-\pi)}{}}((f(z))). \tag{5.19}$$

We recall that in [1826g] this equation was regarded as valid only when $f(z)$ does not have an infinity at $z = 0$, because, under the convention he was using, any residue at $z = 0$ had to be excluded from the sum on the right-hand side. He now remarks that this restriction can be removed, provided that the notation

$$\overset{(R)}{\underset{(0)}{}}\mathscr{E}\overset{(\pi)}{\underset{(-\pi)}{}}((f(z)))$$

is reinterpreted to include the residue (if any) at $z = 0$. He goes to some trouble to show this, arguing as follows. If $f(z)$ has an infinity of order m at $z = 0$, and we write

$$f_0(z) = z^m f(z),$$

$$\varpi(z) = f(z) - \left(\frac{f_0(0)}{z^m} + \frac{1}{z^{m-1}} \frac{f_0'(0)}{1} + \cdots + \frac{1}{z} \cdot \frac{f_0^{(m-1)}(0)}{(m-1)!}\right),$$

then we shall have, with either interpretation of the integral residue,

$$\overset{(R)}{\underset{(0)}{}}\mathscr{E}\overset{(\pi)}{\underset{(-\pi)}{}}((\varpi(z))) = \frac{1}{2\pi}\int_{-\pi}^{\pi} R e^{pi} \varpi(R e^{pi}) \, dp.$$

Substituting the definition of $\varpi(z)$ in both sides of this equation, and observing that

$$\mathscr{E}\left(\left(\frac{1}{z^n}\right)\right) = 0, \quad \int_{-\pi}^{\pi} \frac{R\,e^{pi}}{R^n\,e^{npi}}\,dp = 0 \qquad (n > 1),$$

$$\mathscr{E}\left(\left(\frac{1}{z}\right)\right) = 1, \quad \int_{-\pi}^{\pi} \frac{R\,e^{pi}}{R\,e^{pi}}\,dp = 2\pi,$$

he obtains at once

$$\substack{(R)\\(0)}\mathscr{E}\substack{(\pi)\\(-\pi)}((f(z))) - \frac{f_0^{(m-1)}(0)}{(m-1)!} = \frac{1}{2\pi}\int_{-\pi}^{\pi} R\,e^{pi}f(R\,e^{pi})\,dp - \frac{f_0^{(m-1)}(0)}{(m-1)!},$$

whence

$$\substack{(R)\\(0)}\mathscr{E}\substack{(\pi)\\(-\pi)}((F(z))) = \frac{1}{2\pi}\int_{-\pi}^{\pi} R\,e^{pi}f(R\,e^{pi})\,dp,$$

where he is now using the new interpretation of the integral residue.

Thus Cauchy has here abandoned his special treatment of the centre of the circle as if it were on the boundary. We note that the above argument contains the essential ideas of the usual modern proof of the expression for the residue of a function at a multiple pole.

He now turns to the justification of his assertion that the principal value of $\mathscr{E}((f(z)))$ is the one that should be used in equation (5.17). He supposes that

$$\Delta = zf(z) - F = r\,e^{pi}\,f(r\,e^{pi}) - F$$

tends to zero, for all values of p, when r takes a succession of values tending to infinity. If R is one such value of r and δ is the corresponding value of the integral

$$\frac{1}{2\pi}\int_{-\pi}^{\pi}\Delta\,dp,$$

then it follows from equation (5.19) above that

$$\substack{(R)\\(0)}\mathscr{E}\substack{(\pi)\\(-\pi)}((f(z))) = F + \delta.$$

Letting R become infinite, he obtains the desired equation

$$\mathscr{E}((f(z))) = F, \tag{5.20}$$

where $\mathscr{E}((f(z)))$ is now the principal value of the integral residue.

He goes on to remark that equation (5.20) will still hold under slightly weaker conditions; he now supposes that $\Delta = zf(z) - F$ tends to 0 as z becomes infinite in the way just described, except that there may be a finite number of values of p in the neighbourhood of which Δ is only required to remain finite (in modern language, Δ has to be bounded). To illustrate this situation, he considers the function

$$f(z) = \frac{1}{z}\left(1 + \frac{2}{e^z + e^{-z}}\right).$$

In this case $zf(z) \to 1$ when the modulus r of z goes to infinity through the sequence $r = n\pi$, with n running through the positive integers, except when p is near $\pm\frac{1}{2}\pi$, in which case $zf(z) - 1$ is bounded in absolute value by 1. Equation (5.20) in this case gives the familiar result

$$1 - \tfrac{1}{3} + \tfrac{1}{5} - \tfrac{1}{7} + \cdots = \frac{\pi}{4}.$$

The remainder of [1827d] is devoted to further illustrative examples, including some very elaborate ones. One entertaining result emerges from the equation

$$\mathscr{E}\left(\left(\frac{z}{\sinh az \sinh bz}\right)\right) = 0.$$

He obtains a series transformation that can be written in the form

$$-\frac{1}{a^2}\left\{\frac{1}{\sinh(\pi b/a)} - \frac{2}{\sinh(2\pi b/a)} + \frac{3}{\sinh(3\pi b/a)} - \cdots\right\}$$

$$= \frac{1}{b^2}\left\{\frac{1}{\sinh(\pi a/b)} - \frac{2}{\sinh(2\pi a/b)} + \frac{3}{\sinh(3\pi a/b)} - \cdots\right\} - \frac{1}{2\pi ab}.$$

This result enables him to evaluate the sum

$$\frac{1}{x - (1/x)} - \frac{2}{x^2 - (1/x^2)} + \frac{3}{x^3 - (1/x^3)} - \cdots$$

when x is nearly equal to 1, in which case the series converges extremely slowly. He takes $x = 1\cdot0001$, for which he remarks that one would require about 140 000 terms of the series to calculate its sum to within $0\cdot1$; using the transformation, he is able to calculate the sum to be $2500\cdot124\,997\ldots$. He gives some other series transformations of the same kind.

This paper marks a turning-point in Cauchy's analysis of the relationship between residues and definite integrals. He has recognised that there are valuable advantages to be gained by considering integrals round circles rather than along the sides of rectangles; he has found a simpler and more direct method of evaluating residues at points where a function becomes infinite; and he has removed the anomaly created by his earlier treatment of the centre of a circle as if it were on the boundary. The use of geometrical language has enabled him to improve substantially his handling of the behaviour of a function at infinity. Finally, it is to be noted that here for the first time he has deliberately considered integrals round simple closed curves; for the present they are circles, but even this special case was to earn significant dividends within a few years ([1831d], Section 6.9 below).

5.11. Some further applications of the technique used in [1827d] are given in the immediately following paper [1827e]. Cauchy's main theme here is the expansion of a given function in a series of rational functions.

His main result is as follows. Suppose that, in the notation of [1827d] (Section 5.10 above)

$$f(z) - F = f(r\,e^{pi}) - F$$

tends to 0 when r takes a succession of values tending to infinity for all values of p (or for all but a finite number, in whose neighbourhood the function remains finite). Then we shall have

$$f(x) = \mathcal{E}\,\frac{((f(z)))}{x - z} + F, \qquad (5.21)$$

where the integral residue is required to take its principal value, as it was defined in [1827d].

To prove this, he simply replaces $f(z)$ in equation (5.20) by $f(z)/(x - z)$, and observes that

$$\frac{zf(z)}{z - x}$$

has the limit F as the modulus r of z takes the same succession of values. By the results of [1827d], it follows at once that

$$\mathcal{E}\left(\left(\frac{f(z)}{z - x}\right)\right) = F,$$

i.e.

$$f(x) - \mathcal{E}\,\frac{((f(z)))}{x - z} = F,$$

the required result. In particular, $F = 0$ gives

$$f(x) = \mathcal{E}\,\frac{((f(z)))}{x - z}. \qquad (5.22)$$

He obtains a slight generalisation of this result; if, under the same conditions, $\frac{1}{2}[f(z) + f(-z)]$ has the limit F and $[(f(z) - f(-z)]/2z$ the limit F_1, then

$$f(x) = \mathcal{E}\,\frac{((f(z)))}{x - z} + F + F_1 x.$$

He goes on to show how numerous expansions, including the well-known partial-fraction series for cosec x and sec x, can be derived from these results. It is at this point that he indicates that a result given at the end of [1826d] (Section 5.6 above) is incorrect.

We note here, although Cauchy does not state the theorem explicitly, that it follows immediately from these results that if $f(z)$ has no infinities and has the limit 0 as the modulus r of z becomes infinite, then $f(z)$ is identically 0; in other words, what was to become known as Liouville's theorem is implicit in what he has done here.

In the next note [1827f] Cauchy continues the same investigation from a different point of view. He begins by citing some of Euler's results[7] on series of rational functions; he then raises the question of evaluating the sum of a given infinite series. In the case of a series

$$\sum_{n=1}^{\infty} f(n),$$

where $f(z)$ is an even function, he remarks that the series can also be written as

$$\sum_{n=1}^{\infty} \tfrac{1}{2}[f(n) + f(-n)].$$

Since the zeros of $\sin \pi z$ are $0, \pm 1, \pm 2, \ldots$, it follows that

$$f(0) + f(1) + f(2) + \cdots + f(-1) + f(-2) + \cdots = \mathscr{E} f(z) \frac{\pi \cos \pi z}{((\sin \pi z))},$$

$$(5.23)$$

provided that the series is convergent and $f(0)$ has a finite value. The term $f(0)$ can be dropped from the left-hand side of (5.23) by changing the expression on the right-hand side to

$$\mathscr{E} \frac{f(z)}{z} \left(\left(\frac{\pi z \cos \pi z}{\sin \pi z} \right) \right).$$

If $f(z)$ is a general function, then $f(z) + f(-z)$ will be even, and we shall have an expression for

$$\sum_{n=1}^{\infty} \tfrac{1}{2}[f(n) + f(-n)].$$

He now supposes that $zf(z)$ (or $\tfrac{1}{2}z[f(z) + f(-z)]$) tends to 0 as the modulus r of z tends to infinity through a succession of values, either for all values of z/r or for all but a finite number, in whose neighbourhood the function remains bounded; the same will hold if we introduce a factor

---

[7] Euler [1748], Chapter X and Euler [1785].

$\cos \pi z / \sin \pi z$. By the results of [1827d] (Section 5.10 above) it then follows that

$$\mathscr{E}\left(\left(f(z)\frac{\pi \cos \pi z}{\sin \pi z}\right)\right) = 0. \tag{5.24}$$

Combining (5.23) and (5.24) thus gives

$$f(0) + \sum_{n=1}^{\infty} f(n) + \sum_{n=1}^{\infty} f(-n) = -\mathscr{E}\frac{\pi \cos \pi z}{\sin \pi z}((f(z))),$$

and similarly

$$\sum_{n=1}^{\infty} f(n) + \sum_{n=1}^{\infty} f(-n) = -\mathscr{E}\frac{\pi z \cos \pi z}{\sin \pi z}\left(\left(\frac{f(z)}{z}\right)\right).$$

The remainder of [1827f] is taken up by particular examples of these series transformations. In the first of these he takes

$$f(z) = 1/z^{2m},$$

where m is a positive integer, and arrives at some extremely complicated formulae, related to some of those mentioned in [1827a] (Section 5.9 above) and [1827d] (Section 5.10 above).

In the very brief note [1827g] Cauchy mentions a paper by Euler [1775a] on partial fractions, in which Euler introduces a notion equivalent to the residue of a function $f(z)$ at a simple pole; Euler's technique for determining the partial fraction for a rational function $f(z)/g(z)$ corresponding to a zero $z = a$ of $g(z)$ is to write $z = a + \omega$, and thus expand $f(z)/g(z)$ in powers of ω to find the principal part of $f(z)/g(z)$ near $z = a$. It appears that Cauchy's attention was drawn to this paper of Euler's by Lacroix (Grattan–Guinness [1990], p. 652).

5.12. In the short note [1829b] Cauchy considers for the first time an essential singularity in the finite part of the plane. He begins by remarking that in [1827d] he had tacitly assumed that the number of infinities of $f(z)$ with $|z| \leq R$ is finite for all R; he now proposes to remove this restriction.

He supposes that, if we write as usual

$$z = r(\cos p + \mathrm{i} \sin p),$$

then $zf(z)$ tends to 0 as r runs through a sequence $(\rho, \rho_1, \rho_2, \ldots)$ tending to 0; as in [1827d], this is required to hold for all p, except perhaps for a finite number in whose neighbourhood $zf(z)$ remains bounded. He quotes from [1826g] (Section 5.8 above) the formula

$$\underset{(r_0)}{\overset{(R)}{\mathscr{E}}}\underset{(-\pi)}{\overset{(\pi)}{}}((f(z))) = \frac{1}{2\pi}\int_{-\pi}^{\pi} R\,\mathrm{e}^{p\mathrm{i}}\,f(R\,\mathrm{e}^{p\mathrm{i}})\,\mathrm{d}p - \frac{1}{2\pi}\int_{-\pi}^{\pi} r_0\,\mathrm{e}^{p\mathrm{i}}\,f(r_0\,\mathrm{e}^{p\mathrm{i}})\,\mathrm{d}p.$$

Allowing r_0 to tend to 0 along the given sequence, he concludes that

$$\overset{(R)}{\underset{(0)}{\mathscr{E}}}{}^{(\pi)}_{(-\pi)}((f(z))) = \frac{1}{2\pi} \int_{-\pi}^{\pi} R\, e^{pi}\, f(R\, e^{pi})\, dp, \qquad (5.25)$$

as in [1827d] (Section 5.10 above). We can then examine as before what happens when R becomes infinite.

More generally, if we have

$$f(z) = \frac{A_0}{z^m} + \frac{A_1}{z^{m-1}} + \cdots + \frac{A_{m-1}}{z} + \varpi(z),$$

where $\varpi(z)$ satisfies the conditions imposed above on $f(z)$, then an additional term A_{m-1} has to be added to the left-hand side of equation (5.25).

The remainder of this note is devoted to the examination of illustrative examples, such as

$$f(z) = \frac{\tanh az}{z} \tan\left(\frac{b}{z}\right).$$

We note that the essential singularity considered here is not an isolated one. The first treatment of what happens near an isolated essential singularity seems to have been that given by Laurent [1843].

5.13. In this section we examine some ideas presented by Cauchy on the problem of expressing a given function as a product of linear factors.

In the short note [1829a] in the *Bulletin de Férussac* Cauchy takes a first look at the problem. He remarks that every polynomial and some transcendental functions, such as $\sin x$ and $\cos x$, can be expressed in this way, but the result is not true in general. As an example, he adduces the function

$$f(x) = e^x - 1,$$

whose zeros[8] are at $x = 0$ and at $x = \pm 2n\pi i$, where n is a positive integer; nevertheless, the equation

$$e^x - 1 = x \prod_{n=1}^{\infty} \left(1 + \frac{x}{2n\pi i}\right)\left(1 - \frac{x}{2n\pi i}\right)$$

is false, an additional factor being needed on the right-hand side. The correct formula is then

$$e^x - 1 = x\, e^{x/2} \prod_{n=1}^{\infty} \left(1 + \frac{x}{2n\pi i}\right)\left(1 - \frac{x}{2n\pi i}\right).$$

---

[8] Cauchy gives the zeros as $x = \pm n\pi i$, an obvious slip, which is repeated in *Oeuvres* (2)2, 85.

The additional factor $e^{x/2}$ is not linear, cannot be decomposed into linear factors, and has no zeros.

He therefore proposes to see how much can be achieved in this direction by using the calculus of residues. He then states the following theorem without proof.

Suppose that $f(z)$ and $F(z)$ are finite and continuous and have continuous derivatives of all orders for all finite z, and that $f(0) \neq 0$, $F(0) \neq 0$. Let $\alpha, \beta, \gamma, \ldots$ be the zeros of $f(z)$ and $\lambda, \mu, \nu, \ldots$ those of $F(z)$. Suppose also that the ratio[9]

$$\frac{f'(x/z)F'(z)}{f(x/z)F(z)}$$

tends to 0 as the modulus r of z tends to 0 or to ∞ through appropriate successions of values. Then we shall have

$$\frac{f(x/\lambda)}{f(0)} \frac{f(x/\mu)}{f(0)} \frac{f(x/\nu)}{f(0)} \cdots = \frac{F(x/\alpha)}{F(0)} \frac{F(x/\beta)}{F(0)} \frac{F(x/\gamma)}{F(0)} \cdots . \qquad (5.26)$$

As a first illustration, Cauchy takes $f(z)$ to be a polynomial and $F(z) = \sin(\pi\sqrt{z})/\pi\sqrt{z}$, obtaining

$$\frac{f(x)}{f(0)} \frac{f(x/4)}{f(0)} \frac{f(x/9)}{f(0)} \cdots = \frac{\sin \pi\sqrt{(x/\alpha)}}{\pi\sqrt{(x/\alpha)}} \frac{\sin \pi\sqrt{(x/\beta)}}{\pi\sqrt{(x/\beta)}} \cdots ,$$

where $\alpha, \beta, \ldots$ are the zeros of $f(z)$. Again, the case $f(z) = 1 - z$ leads to the familiar product formula for the sine in the form

$$\frac{\sin(\pi\sqrt{x})}{\pi\sqrt{x}} = (1 - x)\left(1 - \frac{x}{4}\right)\left(1 - \frac{x}{9}\right) \cdots .$$

Other special results include the formula

$$\cos\frac{x}{2} \cos\frac{x}{4} \cos\frac{x}{6} \cdots = \frac{\sin x}{x} \frac{3\sin(x/3)}{x} \frac{5\sin(x/5)}{x} \cdots .$$

In the longer paper [1829c] Cauchy obtains a more general form of the above theorem. He begins with an elementary algebraic proof of equation (5.26) above for the case where $f(z)$ and $F(z)$ are both polynomials. He then gives a residue proof for the same case; since the technique he uses has some points of interest, we give some details of his proof.

Let

$$P = \frac{f(x/\lambda)}{f(0)} \frac{f(x/\mu)}{f(0)} \frac{f(x/\nu)}{f(0)} \cdots ,$$

$$Q = \frac{F(x/\alpha)}{F(0)} \cdot \frac{F(x/\beta)}{F(0)} \cdot \frac{F(x/\gamma)}{F(0)} \cdots ,$$

---

[9] This formula is as given in [1829a], but is wrongly printed in the version in the *Oeuvres*.

where, as before, α, β, γ, ... are the zeros of $f(z)$ and λ, μ, ν, ... are those of $F(z)$. Then

$$\frac{1}{P}\frac{dP}{dx} = \frac{1}{\lambda}\cdot\frac{f'(x/\lambda)}{f(x/\lambda)} + \frac{1}{\mu}\frac{f'(x/\mu)}{f(x/\mu)} + \cdots, \tag{5.27}$$

$$\frac{1}{Q}\frac{dQ}{dx} = \frac{1}{\alpha}\frac{F'(x/\alpha)}{F(x/\alpha)} + \frac{1}{\beta}\frac{F'(x/\beta)}{F(x/\beta)} + \cdots. \tag{5.28}$$

The right-hand sides of (5.27) and (5.28) can then be written as

$$\mathscr{E}\left\{\frac{1}{z}\frac{f'(x/z)}{f(x/z)}\frac{F'(z)}{((F(z)))}\right\}, \qquad \mathscr{E}\left\{\frac{1}{z}\frac{F'(x/z)}{F(x/z)}\frac{f'(z)}{((f(z)))}\right\},$$

respectively. By using the change-of-variable formula for residues from [1826e] (Section 5.7 above) he shows that the second of these expressions is also equal to

$$-\mathscr{E}\left\{\frac{1}{z}\frac{F'(z)}{F(z)}\frac{f'(x/z)}{((f(x/z)))}\right\}.$$

By combining these results, he obtains

$$\frac{1}{P}\frac{dP}{dx} - \frac{1}{Q}\frac{dQ}{dx} = \mathscr{E}\left\{\frac{1}{z}\frac{f'(x/z)}{f(x/z)}\frac{F'(z)}{((F(z)))}\right\} + \mathscr{E}\left\{\frac{1}{z}\frac{F'(z)}{F(z)}\frac{f'(x/z)}{((f(x/z)))}\right\}$$

$$= \mathscr{E}\left(\left(\frac{f'(x/z)F'(z)}{zf(x/z)F(z)}\right)\right).$$

since, as he verifies, the factor z in the denominator of the last expression makes no contribution to the integral residue. Furthermore, since

$$\frac{f'(x/z)}{f(x/z)}\frac{F'(z)}{F(z)}$$

takes the value 0 when z becomes infinite, it follows from the results of [1826a] (Section 5.3 above) that the last integral residue vanishes, so that we have

$$\frac{1}{P}\frac{dP}{dx} - \frac{1}{Q}\frac{dQ}{dx} = 0.$$

It then follows that P and Q are proportional to one another; since $P(0) = Q(0) = 1$, we have $P(x) = Q(x)$ for all x, as required.

Cauchy then proceeds to adapt this argument to the more general case where $f(z)$ and $F(z)$ are not necessarily polynomials. We omit the details of his proof, but give his main result, which is as follows. Suppose that $f(z)$ and $F(z)$ are finite and continuous, together with their derivatives of all orders, for all finite values of z. Let the non-zero roots of the equations

$$f(z) = 0, \quad F(z) = 0,$$

arranged in increasing order of moduli, be $\alpha, \beta, \gamma, \ldots$ for $f(z)$, and $\lambda, \mu, \nu, \ldots$ for $F(z)$. He explicitly assumes that for each function, the number of roots with modulus less than R is finite for all R. Suppose also that

$$\frac{f'(x/z)F'(z)}{f(x/z)F(z)} \left[\text{or} \quad \frac{1}{2} \left\{ \frac{f'(x/z)F'(z)}{f(x/z)F(z)} + \frac{f'(-x/z)F'(-z)}{f(-x/z)F(-z)} \right\} \right]$$

has a limit $G(x)$ as the modulus of z becomes infinite along an appropriate succession of values and that

$$\frac{f'(x/z)F'(z)}{zf(x/z)F(z)}$$

has a certain specified good behaviour as the modulus of z tends to 0 along an appropriate succession of values. Write

$$X(x) = \mathscr{E}\left\{ \frac{f'(x/z)F'(z)}{f(x/z)F(z)} \frac{1}{((z))} \right\}.$$

Then we shall have, for each value ξ of x,

$$\frac{f(x/\lambda)f(x/\mu) \cdots}{f(\xi/\lambda)f(\xi/\mu) \cdots} = \frac{F(x/\alpha)F(x/\beta) \cdots}{F(\xi/\alpha)F(\xi/\beta) \cdots} \times \exp\left\{ \int_{\xi}^{x} [G(x) - X(x)]\,dx \right\}.$$

Cauchy then examines the particular case $f(z) = 1 - z$. Under the assumption that $F(z)$ has a zero of order n at $z = 0$, and that $F'(z)/F(z)$ has the limit G_0 as the modulus of z becomes infinite along an appropriate succession of values, he shows that his result reduces to

$$\frac{F(x)}{F(\xi)} = \frac{1 - (x/\lambda)}{1 - (\xi/\lambda)} \cdot \frac{1 - (x/\mu)}{1 - (\xi/\mu)} \cdots \left(\frac{x}{\xi}\right)^n e^{G_0(x-\xi)}.$$

In particular, if we take $\xi = 0$, we have

$$F(x) = x^n \left(1 - \frac{x}{\lambda}\right)\left(1 - \frac{x}{\mu}\right) \cdots e^{G_0 x} \frac{F^{(n)}(0)}{n!}.$$

He obtains a slightly more general result under the hypotheses that

$$\frac{1}{2}\left\{ \frac{F'(z)}{F(z)} + \frac{F'(-z)}{F(-z)} \right\} \to G_0,$$

$$\frac{1}{2z}\left\{ \frac{F'(z)}{F(z)} - \frac{F'(-z)}{F(-z)} \right\} \to G_1$$

as the modulus of z becomes infinite along a suitable succession of values; here his conclusion is that

$$\frac{F(x)}{F(\xi)} = \frac{1 - (x/\lambda)}{1 - (\xi/\lambda)} \cdot \frac{1 - (x/\mu)}{1 - (\xi/\mu)} \cdots e^{G_0(x-\xi)+\frac{1}{2}G_1(x^2-\xi^2)}.$$

We see that here Cauchy has advanced some distance towards the Weierstrass product representation for entire functions. He gives numerous illustrative examples; in particular, he deals with the case $F(x) = e^x - 1$, which he had adduced in the note [1829a]. We note that, in order to cope with this example, he needs the full force of the more general result described above.

We remark that he has required not only that $f(z)$ and $F(z)$ should be finite and continuous, but also that the same holds for their derivatives of all orders. It is not clear why he should insert this condition here; normally he did not go beyond requiring functions to be 'finite and continuous' until about 1839, and even after that his practice is not consistent.

We note also that he has explicitly imposed the condition that $f(z)$ and $F(z)$ should have at most a finite number of zeros in $|z| \leqslant R$ for all values of R. It was not until many years later that it was realised that this condition is automatically satisfied by any holomorphic function.

5.14. We now try to sum up what Cauchy has achieved in the series of papers discussed in the present chapter.

In [1826a] (Section 5.2 above) he introduces the notion of the residue of a function at a point z_0 where it becomes infinite, defining it as the coefficient of $(z - z_0)^{-1}$ in the expansion of the function about the point. He goes on to define (Section 5.3) the integral residue of a function in a given domain (at this stage either a rectangle or the whole plane) as the sum of its residues at infinities in the domain, and he points out that (in modern language) the integral residue is an additive function of rectangles.

In [1826b] he proves a result equivalent to what has become known as the residue theorem for a rectangle (Section 5.4); this enables him to express the correction term introduced in his earlier work in terms of residues. The result had appeared in disguised form in the 1814 memoir (Chapter 2 above) and more explicitly in his 1817 lectures at the Collège de France (Section 3.3).

In [1826d] he examines what happns to the residues of a function under a change of variable (Section 5.6), and continues this work in [1826e], where he also proves that the residues of the derivative of a single-valued function always vanish (Section 5.7).

In [1826g] he introduces integral residues over more general domains (Section 5.8). His main aim here is to obtain results expressible in polar coordinates; what he gets is essentially the residue theorem for a truncated circular sector. This point of view leads him, in the case where the sector becomes a complete circular disc, to treat the centre of the circle as an exceptional point, almost as if it were on the boundary.

In [1826c] and [1827a] (Sections 5.5 and 5.9) he attempts to improve his conditions on the behaviour of functions at infinity, but on each occasion he tries to apply his results to functions that do not actually satisfy his conditions. It is clear that he is not satisfied with what he has done, and in the important paper [1827d] he adopts a radically new approach (Section 5.10). In this paper he returns to the systematic use of geometrical language for the first time since the 1825 memoir (Chapter 4 above), apart from a casual remark in [1826g] (Section 5.8). What he does is to prescribe the behaviour of the function on a sequence of concentric circles tending to infinity, thus allowing the function to have infinities at points lying between the circles. He also abandons the treatment of the common centre of the circles as an exceptional point. It is clear that he is now thinking for the first time in terms of integrals round closed curves, though at this stage these are only circles. It was not until 1831 (Chapter 6 below) that he began to consider integrals round more general closed curves.

In [1827a] he proves (Section 5.9) the important result that if the zeros of a function $F(z)$ in the unit circle are $z_1, z_1, \ldots, z_m$, and $\varphi(z)F'(z)$ is well behaved there, then

$$\varphi(z_1) + \varphi(z_2) + \cdots + \varphi(z_m) = \frac{1}{2\pi} \int_0^{2\pi} e^{pi} \frac{\varphi(e^{pi})F'(e^{pi})}{F(e^{pi})} \, dp.$$

Formulae like this were to play an important role in Cauchy's work on the Lagrange and other related series (Chapter 6 below).

In [1827e] he uses the technique of [1827d] to expand a given function in a series of rational functions, explicitly exhibiting the infinities of the function (Section 5.11). One immediate consequence of his results here would be Liouville's theorem that a (holomorphic) function bounded in the whole plane is a constant, but he does not state the theorem explicitly.

In the same paper he finds integral expressions for the sums of certain types of infinite series.

In [1829b] he gives serious consideration to a function with a (non-isolated) essential singularity in the finite part of the plane (Section 5.12), and in [1829a] and [1829c] he advances some way towards the Weierstrass product theorem for entire functions (Section 5.13).

From Cauchy's own point of view, what he had created in his calculus of residues was a powerful tool for the evaluation of definite integrals, the summation of series and the discovery of integral expressions for the roots of equations (and the solutions of differential equations, a topic we have ignored). With hindsight, we can see that the most important steps he had made in this series of papers were the switch from rectangular to polar coordinates and his discovery of the value of considering integrals round closed contours.

It has been remarked, e.g., by Freudenthal [1971], p. 139, that there is no explicit reference in these papers to the 1825 memoir (apart from the correction in [1826c] of some special results). On the other hand, some of the techniques of the 1825 memoir do reappear; in [1826a] he uses the idea of separating the singular part of a function, which had played an important role in 1825 (Sections 4.5 and 4.6 above); and in [1827d] he returns to intensive use of the geometrical picture that he had introduced in 1825 (Section 4.8 above). As Cauchy points out in the introduction to the first number of the *Exercices* in 1826, his aim in these papers is to find easier ways of treating problems already solved and to solve new ones. In other words, he is trying to simplify the approach to many of the problems on related topics, instead of referring all the time to very general theorems of the type discussed in the 1825 memoir.

6

The Lagrange series and the Turin memoirs

6.1. In the present chapter we shall be mainly concerned with Cauchy's investigations of the Lagrange series, and his discovery that some of the ideas used there were much more widely applicable; they led him to his important results on the convergence of the Taylor series of an analytic function and on power-series expansions by implicit functions.

We shall begin by outlining some of the early work on the Lagrange series, referring to Lagrange's discovery of the series [1770a] and his application of it to Kepler's problem [1770b], to a paper by Laplace [1779] and the criticism of it by Paoli [1788], and to the first serious investigation of the convergence of the series by Laplace [1825]. We then describe Cauchy's first studies of its convergence in his [1827b] and [1827c], and we go on to the long memoir [1831d], issued in lithograph form during Cauchy's stay in Turin, and containing his results on the convergence of power-series expansions of both explicit and implicit analytic functions, together with his 'calculus of limits', better known today as the method of majorants. The memoir also contains extensive applications of his results to celestial mechanics, but we shall omit any discussion of these. The part of [1831d] with which we shall be concerned first appeared in print (with some small but significant revisions) in Cauchy's *Exercices d'analyse* in 1841; there is also a brief preliminary abstract [1831a], published in the *Bulletin de Férussac*.

We conclude the present chapter with a discussion of Cauchy's second Turin memoir [1831e] (also issued in lithograph form); in this he first considered integrals taken round general simple closed contours in the complex plane, and obtained formulae concerning the zeros of an analytic function inside such a contour; he also constructed a 'calculus of indices', whose results would now be phrased in terms of winding numbers.

Preliminary abstracts of this memoir were published as [1831b] and [1831c] in the *Bulletin de Férussac*.

6.2. The Lagrange series made its first appearance in the course of Lagrange's investigations into methods for the numerical solution of algebraic equations. In Section II of Lagrange's paper [1770a] he begins by considering an equation of the form

$$a - bx + cx^2 - dx^3 + \cdots = 0;$$

he writes the left-hand side in the form

$$a\left(1 - \frac{x}{p}\right)\left(1 - \frac{x}{q}\right)\left(1 - \frac{x}{r}\right)\cdots,$$

so that p, q, r, ... are the roots of the equation. Putting

$$\xi = \frac{1}{b}(cx - dx^2 + ex^3 - \cdots),$$

he obtains an expansion for one of the roots, which he writes in the form

$$p = x + \xi x + \frac{1}{2}\frac{d}{dx}(\xi^2 x^2) + \frac{1}{2 \times 3}\frac{d^2}{dx^2}(\xi^3 x^3) + \cdots,$$

where x is to be replaced by a/b after the differentiations have been performed. He obtains similar expansions for $\log p$, p^2, p^3, ... His proofs are purely formal, his chief tool being the rearrangement of multiple series.

He gives some illustrative examples, including a derivation of Newton's formula for the reversion of series.

He then recasts his results slightly, writing the original equation in the form

$$\alpha - x + \varphi(x) = 0; \tag{6.1}$$

taking p to be a root of (6.1) and $\psi(x)$ to be an arbitrary function, he obtains the expansion

$$\psi(p) = \psi(\alpha) + \varphi(\alpha)\psi'(\alpha) + \frac{1}{2}\frac{d}{d\alpha}[\varphi^2(\alpha)\psi'(\alpha)]$$

$$+ \frac{1}{2 \times 3}\frac{d^2}{d\alpha^2}[\varphi^3(\alpha)\psi'(\alpha)] + \frac{1}{2 \times 3 \times 4}\frac{d^3}{d\alpha^3}[\varphi^4(\alpha)\psi'(\alpha)] + \cdots.$$

It is this series (or minor variants of it) that has become generally known as the Lagrange series.

In later sections of the paper he discusses how to decide which root of (6.1) is represented by the series, and considers methods of finding other roots; he also makes some general remarks about conditions for the series to be convergent.

6.3. In his paper [1770a] Lagrange had used his series for the solution of algebraic equations. In the subsequent paper [1770b] he uses the same method to obtain a solution of Kepler's problem in planetary theory; this was to become the best-known application of the Lagrange series.

To describe Kepler's problem, we take the orbit of a planet to be an ellipse of eccentricity c with the Sun at one focus. The ellipse can be given parametrically by the equations

$$x = a \cos u, \quad y = b \sin u,$$

where a and b are, respectively, the semi-major axis and the semi-minor axis, and are related by the equation

$$b^2 = a^2(1 - c^2);$$

the parameter u is called the *eccentric anomaly*. The *true anomaly* is the angle v between the radius vector from the Sun to the planet and the major axis. The *mean anomaly* t is an angular variable proportional to the time elapsed since the planet was at a fixed point of its orbit (usually its perihelion) and is normalised so that $t = 2\pi$ when the planet has completed one orbit. If the distance from the Sun to the planet is r, it is convenient to call $R = r/a$ the (normalised) *radius vector*. These quantities are usually referred to as the *elements* of the orbit, and the following relations hold between them:

$$t = u - c \sin u,$$

$$R = 1 - c \cos u,$$

$$\tan \tfrac{1}{2}v = \left(\frac{1 + c}{1 - c}\right)^{1/2} \tan \tfrac{1}{2}u.$$

Kepler's equation is then

$$t = u - c \sin u,$$

and *Kepler's problem* is to solve this equation for u in terms of t or to find an expression for a given function $\psi(u)$ of u.

Since the eccentricity c of a planetary orbit is generally quite small, it is natural to try to expand the elements of the orbit or functions of them in series in powers of c. In his paper [1770b] Lagrange obtains these expansions by a formal application of the results of his first paper [1770a], although the equation he wishes to solve is now transcendental rather than algebraic. The expansion he obtains is, apart from some minor changes in notation,

$$\psi(u) = \psi(t) + c \sin t . \psi'(t) + \frac{c^2}{1 \times 2} \frac{d}{dt} [\psi'(t) \sin^2 t]$$

$$+ \frac{c^3}{1 \times 2 \times 3} \frac{d^2}{dt^2} [\psi'(t) \sin^3 t] + \cdots$$

for a given function $\psi(u)$. In particular, we have

$$u = t + c \sin t + \frac{c^2}{1 \times 2} \frac{d}{dt} (\sin^2 t) + \frac{c^3}{1 \times 2 \times 3} \frac{d^2}{dt^2} (\sin^3 t) + \cdots .$$

For the radius vector R, he obtains

$$R = 1 - c \cos t + c^2 \sin^2 t + \frac{c^3}{1 \times 2} \frac{d}{dt} (\sin^3 t)$$

$$+ \frac{c^4}{1 \times 2 \times 3} \frac{d}{dt} (\sin^4 t) + \cdots .$$

He also gives expansions for $\tan \frac{1}{2} v$ and for v itself. By some ingenious manipulations, he obtains expressions for the coefficients of the powers of c in terms of sines and cosines of multiple angles. There is no serious consideration of convergence; towards the end of the paper he simply remarks that the eccentricity has to be very small if the convergence of the series is to be ensured.

6.4. Laplace [1779] proposed a substantial generalisation of the Lagrange series. Given an equation

$$\varphi(x, \alpha) = 0$$

relating the variables x and α, he seeks an expansion for a given function $u(x, \alpha)$ as a series in powers of α. Towards the end of the paper (Section IX) he states the following result without proof. He supposes that $x = a$ is a root of the equation $\varphi(x, 0) = 0$ of multiplicity i, and that $u(x, \alpha)$ has the power-series expansion

$$q_0 + q_1\alpha + q_2\alpha^2 + \cdots ,$$

so that $q_0 = u(a, 0)$; he states that we then have (for $n \geqslant 1$)

$$q_n = \frac{1}{n!} \left[\frac{\partial^n u}{\partial \alpha^n} \right]_{\alpha=0, x=a}$$

$$- \frac{1}{(n-1)!i} \left[\frac{\partial^{n-1}}{\partial x^{n-1}} \left\{ \frac{(x-a)^n}{n!} \left[\frac{\partial^n}{\partial \alpha^n} \left(\frac{\partial u}{\partial x} \log \varphi(x, a) \right) \right]_{\alpha=0} \right\} \right]_{x=a} .$$

In particular, if $\varphi(x, \alpha) = x - a - \alpha z(x)$, and u is given initially as a function of x only, then this expression reduces to

$$q_n = \frac{1}{n!} \left[\frac{\partial^{n-1}}{\partial x^{n-1}} \left(z^n \frac{\partial u}{\partial x} \right) \right]_{x=a} ;$$

in other words, we have the Lagrange series, for which he had already given a formal proof in Section VII.

It was pointed out by Paoli [1788], Sections XI and XII, that this result is incorrect if the multiplicity $i > 1$. If so, and if u is given initially as a function of x only, then there will in general be i roots of $\varphi(x, \alpha) = 0$ that reduce to $x = a$ when $\alpha = 0$; the series will then give not the value of $u(x)$ at one of the roots, but the arithmetic mean of its values at these i roots. Laplace's result could thus be correct only if the root remains of multiplicity i when $\alpha \neq 0$.

This result of Laplace was to be further discussed by Cauchy in his [1831d] (Section 6.11 below).

6.5. Laplace [1825] was the first to investigate seriously the convergence of the Lagrange series; he did so in the context of Kepler's problem (Section 6.3 above). There is a story that when Cauchy insisted at a meeting of the Academy of Sciences that no infinite series should be used until its convergence had been verified, Laplace rushed home to check the convergence of these particular series. Whether or not there is any truth in the story, Laplace did look into the matter; Badolati [1977] suggests that Laplace's paper was actually written in 1823.

In the paper [1825] Laplace begins with the series for the radius vector R, written in the form

$$R = 1 + \tfrac{1}{2}c^2 - c \cos t - \frac{c^2}{2!} \cos 2t + \frac{c^3}{2!2^2} (3 \cos 3t - 3 \cos t) - \cdots ,$$

where c is the eccentricity. The general term of this expansion is

$$(-1)^{n-1} \frac{c^n}{(n-1)!2^{n-1}} \Big[n^{n-2} \cos nt - n(n-2)^{n-2} \cos (n-2)t$$

$$+ \frac{n(n-1)}{2!} (n-4)^{n-2} \cos (n-4)t - \cdots \Big],$$

where the bracketed series is terminated as soon as $n - 2r$ ceases to be positive. Laplace then takes $t = \pi/2$, so that the terms with odd n vanish; for even n he replaces each term by its absolute value. He is thus interested in the convergence of the series whose general term is

$$\frac{c^n}{(n-1)!2^{n-1}}\left[n^{n-2}+n(n-2)^{n-2}+\frac{n(n-1)}{2!}(n-4)^{n-2}\right.$$

$$\left.+\frac{n(n-1)(n-2)}{3!}(n-6)^{n-2}+\cdots\right],$$

the bracketed series being continued only as long as $n>2r$. We omit the details of his estimation; his conclusion is that the series will certainly be convergent if

$$c<\frac{2\sqrt{[\omega(1-\omega)]}}{1-2\omega},$$

where ω is defined by the equation

$$\frac{1-\omega}{\omega}=e^{2/(1-2\omega)}.$$

This gives approximately $\omega=0.08307$, whence the series is convergent if $c<0.66195$.

Laplace reaches the same conclusion for the expansions of the eccentric anomaly and the true anomaly as series in powers of c.

Finally, he considers what happens if the series are rearranged so as to be expressed in terms of sines and cosines of multiples of the mean anomaly, say

$$v=t+a^{(1)}\sin t+a^{(2)}\sin 2t+\cdots+a^{(n)}\sin nt+\cdots,$$

the coefficients $a^{(n)}$ being functions of the eccentricity. He concludes that in this case the series are convergent for all $c<1$. With hindsight we could have predicted this result, since the elements of the orbit are sufficiently smooth functions of the mean anomaly for their Fourier series to be convergent.

6.6. Cauchy's investigation of the convergence of the Lagrange series begin with the two memoirs [1827b] and [1827c], presented to the Académie des Sciences on 9 September 1827, and published[1] in 1829.

An earlier paper by Cauchy [1824] deals only with cases where the Lagrange series breaks off after a finite number of terms, so that no question of convergence arises.

In [1827b] Cauchy's aim is to find an asymptotic expression for the

---

[1] In the table of memoirs in Cauchy's *Oeuvres complètes* **2(15)**, 592, and in the table of publications on p. 601, the correlation between memoirs and publications is inaccurate at this point; P66 and P66*bis* (our [1827b]) should be correlated with M63, and P67 (our [1827c]) with M62, so that M63 should precede M62 in the list of memoirs.

terms of the Lagrange series. He begins with a general discussion of the behaviour for large values of n of integrals of the form

$$S = \int_{x_0}^{x_1} u^n v \, dx,$$

where u and v are real- or complex-valued functions of x; his argument is similar to that given by Laplace [1782]. He supposes that $u'(X) = 0$ for some X in the interval (x_0, x_1); he writes

$$U = u(X), \quad V = v(X), \quad V' = v'(X)$$

and so on. He then puts $u = e^w$ and $w = p + qi$, where p and q are real. Under the hypothesis that $P'' = p''(X) < 0$ and $P = p(X)$ is the maximum value of $p(x)$ in the whole interval (x_0, x_1), Cauchy concludes that

$$S = (1 \pm \varepsilon) \frac{V}{B} \frac{e^{nP} \sqrt{(2\pi)}}{\sqrt{n}} \cos^{1/2} \theta \cdot e^{(nQ - \frac{1}{2}\theta)i}, \tag{6.2}$$

where $B^2 = -P''$, $\tan \theta = Q''/P''$ and $\varepsilon \to 0$ when $n \to \infty$. He then gives some alternative forms of this result, and discusses briefly what happens if $p(x)$ attains its maximum at several points in the interval.

His next move is to apply this result to obtain an asymptotic expression for

$$S_n = \frac{1}{m!} \frac{d^m}{dt^m} \{\varphi(t)[\varpi(t)]^n\}$$

when m and n are both large; the case of interest for the Lagrange series (cf. Section 6.2 above) is that where $m = n - 1$. His main weapon here is his contour-integral formula for derivatives, which he first gave in [1822b] (equation (3.19) in Section 3.7 above); in [1827b] he treats the result as being obvious. Writing $m = \mu n$, he obtains at once

$$S_n = \frac{1}{2\pi} \int_{-\pi}^{\pi} \varphi(t + r\,e^{si}) \left[\frac{\varpi(t + r\,e^{si})}{r^\mu \, e^{s\mu i}} \right]^n ds. \tag{6.3}$$

This is in the right form for the results of the first section to be applied, with

$$u = \frac{\varpi(t + r\,e^{si})}{r^\mu \, e^{s\mu i}}, \quad v = \frac{1}{2\pi} \varphi(t + r\,e^{si}),$$

whence

$$w = p + qi = \log \varpi(t + r\,e^{si}) - \mu \log r - s\mu i,$$

$$w' = \frac{dw}{ds} = p' + q'i = \left[\frac{r\,e^{si}\varpi'(t + r\,e^{si})}{\varpi(t + r\,e^{si})} - \mu \right] i.$$

Using the results of the first section, Cauchy concludes that if (i) s satisfies the equation

$$\frac{r \, e^{si} \varpi'(t + r \, e^{si})}{\varpi(t + r \, e^{si})} = \mu$$

and (ii) the corresponding value of w'' has a negative real part, then the modulus of the factor depending on n in the asymptotic expression for S_n is

$$n^{-1/2} e^{nP} = n^{-1/2} |U|^n,$$

so that the series $\sum S_n$ will be convergent or divergent according as $|U| < 1$ or $|U| > 1$.

To simplify the statement of this result, Cauchy writes[2]

$$\psi(x) = \frac{\varpi(t + x)}{x^{\mu}};$$

one then has to seek a root $x = \rho \, e^{si}$ of the equation $\psi'(x) = 0$ such that $R = |\psi(\rho \, e^{si})|$ is a maximum as s varies in $(0, 2\pi)$. The number R he calls the *principal modulus* of the function $\psi(x)$. His conclusion, which he states as a formal theorem, is that if[3]

$$S_n = \frac{1}{m!} \frac{d^m}{dt^m} \{\varphi(t)[\varpi(t)]^m\},$$

and $\mu = m/n$ is finite and positive, then $\sum S_n$ is convergent or divergent according as $R < 1$ or $R > 1$.

He specialises this result to the case $m = n - 1$, remarking that we can then take $\mu = 1$, since $\mu = (n - 1)/n$ is practically equal to 1 when n is large; in this case S_n/n is the general term of the Lagrange series for a value of $\int \varphi(z) \, dz$ at a root z of the equation

$$z = t + \varpi(z).$$

The remainder of [1827b] is devoted to special cases and illustrative examples. In particular, he examines Kepler's equation

$$z = t + c \sin z,$$

where z is the eccentric anomaly, t the mean anomaly, and c the eccentricity of the orbit. His result agrees with Laplace's (Section 6.5 above) for the case $t = \pi/2$; he does some work on general values of t without reaching a final result, but he does show that in the case $t = 0$ the series is convergent for $c < 1$.

In this paper Cauchy has discovered that the expressions for the higher derivatives of an analytic function in terms of contour integrals can be made into a powerful tool. In the present case he has used it to determine the asymptotic behaviour of an expression involving these derivatives, but

---

[2] There is a misprint in equation (9) in the original paper, repeated in the *Oeuvres complètes*; x^r should read x^{μ}.

[3] There is a misprint in equation (12) of the original paper, repeated in the *Oeuvres*; d^n/dt^n should read d^m/dt^m.

he was soon (Sections 6.8 and 6.9 below) to find that it could be used to estimate such expressions directly without having an asymptotic expansion at his disposal.

6.7. In [1827c], the second paper of this group, Cauchy's main aim is to obtain an explicit expression for the remainder in the Lagrange series.

He starts from the formula

$$f(\zeta) = \mathscr{E} \frac{1 - h\varpi'(z)}{((z - x - h\varpi(z)))} f(z),$$

where the sign $\mathscr{E}$ indicates that he is taking the residue at a root ζ (tacitly assumed simple) of the equation

$$z - x - h\varpi(z) = 0.$$

He expands

$$\frac{1 - h\varpi'(z)}{z - x - h\varpi(z)}$$

in powers of h with an explicit remainder term. After a good deal of manipulation, whose details we omit, he arrives at the equation

$$\frac{1 - h\varpi'(z)}{z - x - h\varpi(z)} f(z) = \frac{f(z)}{z - x} - \frac{h}{1} f(z) \frac{\partial}{\partial z} \left(\frac{\varpi(z)}{z - x} \right)$$

$$- \frac{h^2}{2} f(z) \frac{\partial}{\partial z} \left(\frac{\varpi(z)}{z - x} \right)^2 - \cdots$$

$$- \frac{h^n}{n} f(z) \frac{\partial}{\partial z} \left(\frac{\varpi(z)}{z - x} \right)^n + \frac{h^{n+1}\chi(z)}{(z - \zeta)(z - x)^{n+1}}, \quad (6.4)$$

where

$$\frac{\chi(z)}{z - \zeta} = \frac{[\varpi(z)]^n [\varpi(z) - (z - x)\varpi'(z)]}{z - x - h\varpi(z)} f(z).$$

He then takes residues of the two sides of equation (6.4) at $z = x$ and $z = \zeta$, and uses the formula

$$\mathscr{E}((u(z)v'(z) + u'(z)v(z))) = 0,$$

proved in [1826e] (Section 5.7 above); he thus obtains[4]

$$f(\zeta) = f(x) + hf'(x)\varpi(x) + \frac{h^2}{2} \frac{d}{dx} [f'(x)(\varpi(x))^2] + \cdots$$

$$+ \frac{h^n}{n!} \frac{d^{n-1}}{dx^{n-1}} [f'(x)(\varpi(x))^n] + \mathscr{E} \frac{h^{n+1}\chi(z)}{(((z - \xi)(z - x)^{n+1}))}.$$

---

[4] There is a misprint in equation (10) of the original paper, repeated in the *Oeuvres*; the term $(h/1)f(x)\varpi(x)$ on the right-hand side should read $(h/1)f'(x)\varpi(x)$.

He observes that what we have here is just the Lagrange series, with a remainder term

$$r_n = \mathscr{E} \frac{h^n \psi(z)}{(((z - \zeta)(z - x)^n))},$$
(6.5)

where

$$\frac{\psi(z)}{z - \zeta} = \frac{[\varpi(z)]^{n-1}[\varpi(z) - (z - x)\varpi'(z)]}{z - x - h\varpi(z)} f(z).$$

In the next section he transforms his expression for r_n into a definite integral; to achieve this he writes down a Taylor series for $\psi(z)$, with a remainder, in the form

$$\psi(z) = \psi(x) + (z - x)\psi'(x) + \cdots + \frac{(z - x)^{n-1}}{(n - 1)!}\psi^{(n-1)}(x)$$

$$+ \frac{(z - x)^n}{(n - 1)!} \int_0^1 u^{n-1} \psi^{(n)}[z - u(z - x)] \, du,$$

and substitutes this expression in (6.5), obtaining

$$r_n = \frac{h^n}{(n - 1)!} \int_0^1 u^{n-1} \psi^{(n)}[\zeta - u(\zeta - x)] \, du.$$
(6.6)

The rest of [1827c] is concerned with further manipulations and with special cases; these are not of interest for our purposes.

It does not appear that Cauchy ever found any use for these formulae; the chief interest of the paper lies in its indication that Cauchy was still looking for a simple way of investigating the convergence of the Lagrange series. The approach used here was to be superseded by the much more general theory described in the first Turin memoir [1831d] (Sections 6.8–6.13 below).

We may also remark that there is little in this paper that involves applications of complex function theory, merely the use of residue notation and of the equation

$$\mathscr{E}((u(z)v'(z) + u'(z)v(z))) = 0.$$

6.8. Cauchy exiled himself from France, for political reasons, after the 1830 revolution, and spent a few years as professor of theoretical physics at the University of Turin. In October and November 1831 he submitted the two important memoirs [1831d] and [1831e] to the Royal Academy of Sciences at Turin.

The memoir [1831d] was issued at Turin in a lithographed version in 1833. It consists of three main parts. Part I contains a short section on the variation of arbitrary constants (pp. 1–4), a long section entitled 'Formules pour le développement des fonctions en séries, calcul des limites' (pp. 5–56) and a short section on the expansion of $(a + x)^i$ in an infinite series (pp. 57–59). The long Part II is devoted to applications of the results of Part I to celestial mechanics (pp. 60–152): there is also an 'Addition' (pp. 153–209) containing miscellaneous further results. In 1834 an Italian translation by P. Frisiani and G. Piola of Parts I and II of the memoir was published in volume 2 of the *Opuscoli Matematici e Fisici* at Milan; the translators also provided some 82 pages of detailed commentary.

We shall be interested in pp. 5–56 of the memoir, i.e. the principal section of Part I. This was reprinted in Cauchy's *Exercices d'analyse et de physique mathématique* in 1841, with some minor but important modifications. A preliminary abstract of the memoir was published as [1831a] in the *Bulletin de Férussac*, and was reproduced (with an additional footnote) as a preface to the printed version of 1841.

The main theme of the part of [1831d] with which we shall be concerned is the determination of conditions for the expansion of explicit and implicit functions in power series; in the course of this work Cauchy developed methods for the estimation of remainders in such expansions after an arbitrary finite number of terms. He gave this technique the name 'calcul des limites', it is better known today as the 'method of majorants'.

In the preliminary abstract [1831a], Cauchy insists on the necessity of proving the convergence of the most important infinite series arising in celestial mechanics. He then describes his results on the convergence of the Taylor series of a given function and of power-series expansions of implicit functions, and on the estimation of remainders in these expansions. He also describes in general terms the applications of his results in celestial mechanics, and mentions his intention of publishing similar results on the solutions of differential equations.

6.9. The part of [1831d] with which we are concerned begins with the remark that if n is a positive integer, then

$$\int_{-\pi}^{\pi} e^{npi} \, dp = \int_{-\pi}^{\pi} e^{-npi} \, dp = 0.$$

On the other hand, the case $n = 0$ gives

$$\int_{-\pi}^{\pi} dp = 2\pi.$$

It follows at once that, if $f(x)$ is a polynomial, say

$$f(x) = a_0 + a_1 x + \cdots + a_n x^n,$$

and[5] $\xi = X e^{pi}$, where X is the modulus of ξ, then

$$\int_{-\pi}^{\pi} f(\xi)\,dp = \int_{-\pi}^{\pi} f(1/\xi)\,dp = 2\pi a_0 = 2\pi f(0).$$

We note that Cauchy is here integrating round a closed curve (in this case, a circle); he had first deliberately done so in [1827d] (Section 5.10 above).

Cauchy's next aim is to extend this result to a general function $f(x)$, assumed to be finite and continuous,[6] say for $|x| \leqslant X$. He starts from the equation

$$D_X f(\xi) = \frac{1}{X i} D_p f(\xi),$$

and integrates each side with respect to X from 0 to X and with respect to p from $-\pi$ to π; the left-hand side gives

$$\int_{-\pi}^{\pi} dp \int_0^X D_X f(\xi)\,dX = \int_{-\pi}^{\pi} [f(\xi) - f(0)]\,dp$$

$$= \int_{-\pi}^{\pi} f(\xi)\,dp - 2\pi f(0),$$

and the right-hand side gives

$$\int_0^X dX \int_{-\pi}^{\pi} \frac{1}{X i} D_p f(\xi)\,dp = 0.$$

The required result

$$\int_{-\pi}^{\pi} f(\xi)\,dp = 2\pi f(0) \tag{6.7}$$

then follows at once.

---

[5] For ease of printing, we shall write ξ instead of Cauchy's $\bar{x}$.

[6] In the version of [1831d] published in his *Exercices d'Analyse* in 1841, Cauchy makes the additional assumption that $f(x)$ has a continuous derivative; he first introduced this hypothesis in [1839], remarking there that one cannot be certain that a function and its derivative necessarily have their infinities and discontinuities at the same values of the independent variable, although this is true for the examples that turn up in practice. It is clear that at this stage Cauchy still believed that a continuous function, even of a complex variable, has a derivative in general, although there may be isolated points where the derivative becomes infinite or discontinuous.

Frisiani and Piola, in Note II to their translation, say that the author imposes the condition [of continuity] on $f(\xi)$ rather than on the derivative $f'(\xi)$, because usually (but not always) the former condition implies the latter. In Note IV, they remark that Cauchy seems to regard a function that becomes many-valued or indeterminate for some value of ξ, such as

$$\sqrt{(1+\xi)}, \quad \frac{\xi}{1 + \sqrt{(1+\xi^2)}}, \quad \cos(1/\xi),$$

as being discontinuous; their point is that the arguments that Cauchy uses in his proof would fail for such functions, because their derivatives become infinite for certain values of ξ.

The argument used here has some similarity with one used by Gauss [1816] in his third proof of the fundamental theorem of algebra.

Repeated integration by parts with respect to p then gives

$$\int_{-\pi}^{\pi} \frac{f(\xi)}{\xi^n} \, dp = \frac{2\pi f^{(n)}(0)}{n!}, \tag{6.8}$$

provided that $f'(\xi)$, $f''(\xi)$, $\ldots$, $f^{(n)}(\xi)$ are finite and continuous for $|\xi| \leqslant X$.

In particular, if $f(0) = 0$, it follows from (6.7) that

$$\int_{-\pi}^{\pi} f(\xi) \, dp = 0. \tag{6.9}$$

Cauchy then replaces $f(\xi)$ in (6.9) by

$$\frac{\xi[f(\xi) - f(x)]}{\xi - x},$$

where $|x| < X$, so that $x \neq \xi$, and so obtains

$$\int_{-\pi}^{\pi} \frac{\xi[f(\xi) - f(x)]}{\xi - x} \, dp = 0,$$

whence

$$\int_{-\pi}^{\pi} \frac{\xi f(\xi)}{\xi - x} \, dp = f(x) \int_{-\pi}^{\pi} \frac{\xi}{\xi - x} \, dp$$

$$= f(x) \int_{-\pi}^{\pi} \left(1 + \frac{x}{\xi} + \frac{x^2}{\xi^2} + \cdots \right) dp$$

$$= 2\pi f(x). \tag{6.10}$$

This result is essentially what was to become known as 'Cauchy's formula' for the case of a circle; his proof is by far the neatest that he had given up to this time. An equivalent formula appears in [1822b] (Section 3.7 above, equation (3.22)), but he was not at that time thinking in terms of integrals taken around closed curves (cf. Section 5.10 above).

Cauchy now converts (6.10) into an infinite series, observing that if we substitute

$$\frac{\xi}{\xi - x} = 1 + \frac{x}{\xi} + \frac{x^2}{\xi^2} + \cdots$$

into the left-hand side of the equation, we obtain

$$f(x) = \frac{1}{2\pi} \left\{ \int_{-\pi}^{\pi} f(\xi) \, dp + x \int_{-\pi}^{\pi} \frac{f(\xi)}{\xi} \, dp + x^2 \int_{-\pi}^{\pi} \frac{f(\xi)}{\xi^2} \, dp + \cdots \right\}. \tag{6.11}$$

He goes on to remark that the power-series expansion (6.11) will be valid provided that the modulus of x is less than the least value at which $f(x)$ ceases to be finite and continuous. In other words, he has obtained the fundamental result that the series is convergent up to the singularity of $f(x)$ nearest to $x = 0$.

He gives some examples: for instance, it follows from his result that functions such as

$$\cos x, \ \sin x, \ e^x, \ e^{x^2}, \ \cos(1 - x^2)$$

can be expanded as convergent power series for all x; functions such as

$$\sqrt{(1 + x)}, \ 1/(1 - x), \ x/[1 + \sqrt{(1 - x^2)}]$$

have expansions convergent for $|x| < 1$, which may be, and actually are, divergent for $|x| > 1$; and functions such as

$$e^{1/x}, \ e^{1/x^2}, \ \cos(1/x),$$

being discontinuous at $x = 0$, have no expansion in powers of x.

He finally remarks that, by equation (6.8) above, the expansion (6.11) can also be written as

$$f(x) = f(0) + \frac{x}{1}f'(0) + \frac{x^2}{2!}f''(0) + \cdots, \tag{6.12}$$

which is precisely the familiar Maclaurin series.

Cauchy's next aim is to obtain estimates for the general term of the series and for the remainder after n terms. He denotes the maximum modulus of $f(\xi)$ under the condition $|\xi| = X$ by $\wedge f(\xi)$; in modern notation,

$$\wedge f(\xi) = \sup\{|f(\xi)| : |\xi| = X\}.$$

He then argues[7] that

$$\left|\frac{x^n}{n!}f^{(n)}(0)\right| = \left|\frac{x^n}{2\pi}\int_{-\pi}^{\pi}\frac{f(\xi)}{\xi^n}\,dp\right|$$

$$\leqslant \left(\frac{|x|}{X}\right)^n \wedge f(\xi). \tag{6.13}$$

This is essentially the result that came to be known as 'Cauchy's inequality'.

---

[7] In a footnote at this point he sketches a proof of the inequality (in modern notation)

$$\left|\int_a^b \varpi(x)\,dx\right| \leqslant (b - a)\sup\{|\varpi(x)| : a \leqslant x \leqslant b\}.$$

Frisiani and Piola give a detailed proof of this in Note I of their commentary.

Since the right-hand side of (6.13) is the general term in the expansion of

$$\frac{X}{X - |x|} \wedge f(\xi)$$

as a geometric series, it also follows that the modulus of the remainder after n terms of the series (6.12) cannot exceed

$$\frac{|x|^n}{X^{n-1}(X - |x|)} \wedge f(\xi). \tag{6.14}$$

Cauchy also obtains an explicit expression for the remainder by using the identity

$$\frac{\xi}{\xi - x} = 1 + \frac{x}{\xi} + \frac{x^2}{\xi^2} + \cdots + \frac{x^{n-1}}{\xi^{n-1}} + \frac{x^n}{\xi^{n-1}(\xi - x)}.$$

Substituting this expression in (6.10) above, and using equation (6.8), he obtains

$$f(x) = f(0) + \frac{x}{1}f'(0) + \cdots + \frac{x^{n-1}}{(n-1)!}f^{(n-1)}(0)$$

$$+ \frac{1}{2\pi}\int_{-\pi}^{\pi}\frac{x^n}{\xi^{n-1}(\xi - x)}f(\xi)\,\mathrm{d}p. \tag{6.15}$$

From (6.15) we can again deduce that the modulus of the remainder cannot exceed the expression (6.14).

At this point Cauchy sums up all these results in a formal theorem[8] (his Theorem I).

He next gives (without explicit proofs) an extension of these results to functions of several variables. His main result[9] (his Theorem II) is that a function $f(x, y, z, \ldots)$ can be expanded as a multiple power series in $(x, y, z, \ldots)$ if it is finite and continuous[10] for $|x| \leqslant X$, $|y| \leqslant Y$, $|z| \leqslant Z$, $\ldots$; the general term of the series and the remainder after a finite number of terms will not exceed the corresponding quantities for the series whose sum is

$$\frac{X}{X - |x|} \cdot \frac{Y}{Y - |y|} \cdot \frac{Z}{Z - |z|} \cdots \wedge f(\xi, \eta, \zeta, \ldots),$$

where $|\xi| = X, |\eta| = Y, |\zeta| = Z, \ldots$.

---

[8] [1831d], p. 9; *Exercices d'analyse*, **2**, 54; *Oeuvres* **(2)12**, 64.
[9] [1831d], p. 10; *Exercices d'analyse*, **2**, 55; *Oeuvres* **(2)12**, 64–5.
[10] In the 1841 version he adds the condition that the first-order derivatives of $f(x, y, z, \ldots)$ with respect to $x, y, z, \ldots$ should also be finite and continuous.

He remarks afterwards that this result could also be derived from his Theorem I by considering the expansion of $f(\alpha x, \alpha y, \alpha z, \ldots)$ in powers of α and thus taking $\alpha = 1$.

In subsequent paragraphs, Cauchy discusses techniques for finding expressions or upper estimates for $\wedge f(\xi)$ or $\wedge f(x, y, z, \ldots)$, and gives some illustrative examples; he also discusses briefly the expression

$$\wedge' f(\xi) = \min \{|f(\xi)|: |\xi| = X\}$$

(in modern notation).

He concludes this part of the paper with a technical remark: since the modulus of the remainder, when expressed in the form

$$\left| \frac{1}{2\pi} \int_{-\pi}^{\pi} \frac{x^n}{\xi^{n-1}(\xi - x)} f(\xi) \, \mathrm{d}p \right|$$

(equation (6.15) above) will not exceed

$$\wedge \left\{ \frac{x^n}{\xi^{n-1}(\xi - x)} f(\xi) \right\},$$

it is conceivable that one might get a better estimate by using this expression instead of

$$\frac{|x|^n}{X^{n-1}(X - |x|)} \wedge f(\xi).$$

He points out, however, that this procedure would tend to involve one in difficult calculations, whereas in practice the cruder estimate is easier to use and gives all one needs. He also remarks that it is advantageous, when one is using either of these expressions, to choose X in such a way as to give the estimate its smallest possible value; this leads to what he had described in [1827b] (Section 6.6 above) as the 'principal modulus' of the expression. He shows how this procedure works for some special functions, and discusses some particular cases of his general results.

We can now sum up what Cauchy has achieved in this portion of the memoir [1831d]. He has established the convergence of the Maclaurin series of a function up to the singularity nearest to the origin; he has obtained estimates for the individual terms of the series and for its remainder after a finite number of terms, and, in particular, he has established 'Cauchy's inequality'

$$\frac{|f^{(n)}(0)|}{n!} \leqslant \frac{\wedge f(\xi)}{|\xi|^n}.$$

These estimates constitute the central core of what Cauchy was to describe as the 'calcul des limites', later to be known as the 'method of majorants'.

6.10. In the next portion of [1831d] Cauchy applies his basic results to obtain integral expressions for solutions of equations.

He begins by remarking that the integral formulae he has obtained (Section 6.9 above) may be regarded as special cases of the equation[11]

$$\int_{-\pi}^{\pi} \varphi(\xi)\, dp = 2\pi \, {}^{(X)}_{(0)}\mathscr{E}^{(\pi)}_{(-\pi)} \left(\left(\frac{\varphi(x)}{x} \right) \right)$$

of [1827d] (Section 5.10 above), holding, he says, for functions $\varphi(x)$ having a unique determinate value for $0 \leqslant |x| \leqslant X$.

He now supposes that an implicit function y of x is defined by an equation

$$f(x, y) = 0. \tag{6.16}$$

Concentrating on the neighbourhood of a particular value b of y, he writes $y = b + z$, so that (6.16) becomes

$$f(x, b + z) = 0. \tag{6.17}$$

He then assumes that the value of x (fixed for the present) is so chosen that (6.17) has a unique root z such that $|z| < Z$, say; his aim is to find an explicit expression for a given function $F(y) = F(b + z)$ valid for small values of z. Assuming that $F(b + z)$ is 'finite and continuous' for $|z| \leqslant Z$, he writes

$$\chi(x, y) = D_y f(x, y),$$

where D_y denotes partial derivation with respect to y.

Since

$$\frac{\chi(x, b + \zeta)}{f(x, b + \zeta)}$$

has a simple infinity at $\zeta = z$, with residue 1, we have at once[12]

$$F(y) = {}^{(Z)}_{(0)}\mathscr{E}^{(\pi)}_{(-\pi)} \left(\left(\frac{\chi(x, b + \zeta)}{f(x, b + \zeta)} F(b + z) \right) \right)_z.$$

Writing $\zeta = Z\, e^{qi}$ when $|\zeta| = Z$, Cauchy deduces from the residue formula for a circle that

$$F(y) = \frac{1}{2\pi} \int_{-\pi}^{\pi} \zeta \frac{\chi(x, b + \zeta)}{f(x, b + \zeta)} F(b + \zeta)\, dq. \tag{6.18}$$

---

[11] In the 1841 printed version in the *Exercices d'analyse*, the double brackets of his residue notation are replaced by a single angled bracket; this notation was not used in the original lithograph or in the Italian translation. For consistency, we shall continue to use his original double-bracket notation.

[12] The subscript z on the double bracket indicates that the residue is to be taken with respect to the variable z. This notation first appears in the 1841 printed version; we shall use it on occasions when it is desirable to avoid ambiguity.

He gives an alternative proof of (6.18) by using the device introduced in the 1825 memoir and attributed in Chapter 4 above to Ostrogradskiĭ. He removes the singularity by considering

$$\zeta \left\{ \frac{\chi(x,\, b + \zeta)}{f(x,\, b + \zeta)} F(b + \zeta) - \frac{F(b + z)}{\zeta - z} \right\},$$

which has a finite limit when ζ approaches z, and hence has residue 0 there; consequently

$$\int_{-\pi}^{\pi} \zeta \frac{\chi(x,\, b + \zeta)}{f(x,\, b + \zeta)} F(b + \zeta)\, dq = F(b + z) \int_{-\pi}^{\pi} \frac{\zeta\, dq}{\zeta - z}$$

$$= 2\pi F(b + z)$$

$$= 2\pi F(y)$$

as required.

Taking $F(y) = 1$ thus gives

$$\frac{1}{2\pi} \int_{-\pi}^{\pi} \zeta \frac{\chi(x,\, b + \zeta)}{f(x,\, b + \zeta)}\, dq = 1,$$

and $F(y) = y$ gives

$$y = \frac{1}{2\pi} \int_{-\pi}^{\pi} \zeta(b + \zeta) \frac{\chi(x,\, b + \zeta)}{f(x,\, b + \zeta)}\, dq$$

$$= b + \frac{1}{2\pi} \int_{-\pi}^{\pi} \zeta^2 \frac{\chi(x,\, b + \zeta)}{f(x,\, b + \zeta)}\, dq,$$

so that we have here an explicit formula for the root $y = b + z$ of the equation

$$f(x,\, y) = f(x,\, b + z) = 0.$$

Cauchy now goes on to the more general case where the equation $f(x,\, b + z) = 0$ has m roots $z,\, z_1,\, \ldots,\, z_{m-1}$ with $|z_i| < Z$ (multiple roots being counted according to their multiplicity). Following the same line of argument as for $m = 1$ and writing $y,\, y_1,\, \ldots,\, y_{m-1}$ for the corresponding roots of $f(x,\, y) = 0$, he obtains the formula

$$F(y) + F(y_1) + \cdots + F(y_{m-1}) = \overset{(Z)}{\underset{(0)}{\mathscr{E}}} \overset{(\pi)}{\underset{(-\pi)}{\mathscr{E}}} \left(\left(\frac{\chi(x,\, b + z)}{f(x,\, b + z)} F(b + z) \right) \right)_{z}$$

$$= \frac{1}{2\pi} \int_{-\pi}^{\pi} \zeta \frac{\chi(x,\, b + \zeta)}{f(x,\, b + \zeta)} F(b + \zeta)\, dq.$$

$$(6.19)$$

The particular cases $F(y) = 1$ and $F(y) = y$ then give

$$m = \frac{1}{2\pi} \int_{-\pi}^{\pi} \zeta \frac{\chi(x, b+\zeta)}{f(x, b+\zeta)} \, dq, \tag{6.20}$$

$$y_1 + y_2 + \cdots + y_m = mb + \frac{1}{2\pi} \int_{-\pi}^{\pi} \zeta^2 \frac{\chi(x, b+\zeta)}{f(x, b+\zeta)} \, dq. \tag{6.21}$$

We see that (6.20) is essentially the well-known formula for the number of zeros of an analytic function inside a circle; he develops this result further in the next part of the memoir (Section 6.11 below). The result (6.21) confirms the criticism made by Paoli [1788] of Laplace's statement in his [1779] (cf. Section 6.4 above).

6.11. Up to this point Cauchy has been considering the roots y, $y_1, \ldots, y_{m-1}$ of the equation $f(x, y) = 0$ for a fixed value of x, and has obtained formulae for expressions of the form

$$F(y) + F(y_1) + \cdots + F(y_{m-1}). \tag{6.22}$$

He now allows x to vary, so that the roots become functions of x, and he aims at finding expansions for expressions such as (6.22) as power-series in x.

His first result[13] (his Theorem III) gives conditions for the number m of roots to remain unchanged as x varies, and is as follows. Suppose that (i) the equation $f(0, b+z) = 0$ has exactly m roots z with $|z| < Z$, (ii) that $f(x, b+z)$ is finite and single-valued when $|x| < X$ and $|z| < Z$, and (iii) that, when $|x| < X$, the expression

$$\log \frac{f(x, b+\zeta)}{f(0, b+\zeta)},$$

where $|\zeta| = Z$, can be expanded as a convergent power-series in x, say

$$\log \frac{f(x, b+\zeta)}{f(0, b+\zeta)} = xu_1(\zeta) + x^2 u_2(\zeta) + x^3 u_3(\zeta) + \cdots; \tag{6.23}$$

then, for all x with $|x| < X$, the equation

$$f(x, b+z) = 0$$

also has exactly m roots z with $|z| < Z$.

To prove this, he differentiates (6.23) with respect to ζ, obtaining

$$\frac{\chi(x, b+\zeta)}{f(x, b+\zeta)} = \frac{\chi(0, b+\zeta)}{f(0, b+\zeta)} + xu_1'(\zeta) + x^2 u_2'(\zeta) + \cdots, \tag{6.24}$$

---

[13]   [1831d], p. 23; *Exercices d'analyse*, **2**, 66; *Oeuvres* **(2)12**, 78.

where, as before (Section 6.10 above),

$$\chi(x, b + \zeta) = D_\zeta f(x, b + \zeta).$$

Writing $\zeta = Z e^{qi}$, he multiplies (6.24) by ζ and integrates with respect to q from $-\pi$ to π. Since, by [1826e] (Section 5.7 above),

$$\int u'_n(\zeta) \, d\zeta = 0 \qquad (n = 1, 2, \ldots),$$

the integral being taken round the circle $|\zeta| = Z$, we are left with the equation

$$\int_{-\pi}^{\pi} \zeta \frac{\chi(x, b + \zeta)}{f(x, b + \zeta)} \, dq = \int_{-\pi}^{\pi} \zeta \frac{\chi(0, b + \zeta)}{f(0, b + \zeta)} \, d\zeta. \tag{6.25}$$

By the results described in Section 6.10 above, the right-hand side of (6.25) has the value $2\pi m$; the desired conclusion follows.

We note that, as usual, Cauchy has taken for granted the validity of term-by-term differentiation and integration of infinite series. What is more interesting is his introduction of the condition that there exists a power-series expansion of the form (6.23); it is unusual for him to take such a condition as a hypothesis, although he had done so in [1823b] (Section 3.12 above).

Cauchy then gives a corollary[14] (his Theorem IV) to the effect that it is sufficient to assume that

$$\frac{\chi(x, b + \zeta)}{f(x, b + \zeta)}$$

can be expanded as a power-series in x, justifying this statement by saying that it is this assumption that he has used in the proof of his Theorem III. In this case, however, it is not obvious that the integrals of the terms in (6.24) round the circle $|\zeta| = Z$ will necessarily vanish.

Cauchy now turns to his main problem, that of finding power-series expansions for sums of the form

$$F(y) + F(y_1) + \cdots + F(y_{m-1}),$$

where the function $F(b + z)$ is assumed to be 'finite and continuous' for $|z| < Z$. To do this, he multiplies (6.24) by $\zeta F(b + \zeta)$ and integrates with respect to q from $-\pi$ to π. Writing $v_n(\zeta) = u'_n(\zeta)$ and denoting the roots of the equation $f(0, y) = 0$ such that $|y - b| = |z| < Z$ by $\beta, \beta_1, \ldots, \beta_{m-1}$, he obtains

---

[14] [1831d], p. 25; *Exercices d'analyse*, **2**, 68; *Oeuvres* (2)**12**, 80.

$$F(y) + F(y_1) + \cdots + F(y_{m-1}) = F(\beta) + F(\beta_1) + \cdots + F(\beta_{m-1})$$

$$+ \frac{x}{2\pi} \int_{-\pi}^{\pi} \zeta v_1(\zeta) F(b + \zeta) \, dq$$

$$+ \frac{x^2}{2\pi} \int_{-\pi}^{\pi} \zeta v_2(\zeta) F(b + \zeta) \, dq + \cdots$$

$$(6.26)$$

$$= F(\beta) + F(\beta_1) + \cdots + F(\beta_{m-1})$$

$$- \frac{x}{2\pi} \int_{-\pi}^{\pi} \zeta u_1(\zeta) F'(b + \zeta) \, dq$$

$$- \frac{x^2}{2\pi} \int_{-\pi}^{\pi} \zeta u_2(\zeta) F'(b + \zeta) \, dq - \cdots,$$

$$(6.27)$$

the second equation being arrived at by integration by parts.

He writes out in detail what these equations become when $F(y) = y$ and when $m = 1$, $\beta = b$. He also notes that one can obtain an immediate estimate for the terms of (6.26); the modulus of the coefficient of x^n cannot exceed the maximum modulus, for $|\zeta| = Z$, of $\zeta v_n(\zeta) F(b + \zeta)$.

At this point Cauchy introduces a further generalisation, with the remark that most of the results he has obtained continue to hold if $F(y)$ is replaced by a function $F(x, y) = F(x, b + z)$, assumed to be 'finite and continuous' for $|x| < X$ and $|z| < Z$. The basic equation (6.18) of Section 6.10 now becomes

$$F(x, y) + F(x, y_1) + \cdots + F(x, y_{m-1})$$

$$= \frac{1}{2\pi} \int_{-\pi}^{\pi} \zeta \frac{\chi(x, b + \zeta)}{f(x, b + \zeta)} F(x, b + \zeta) \, dq, \qquad (6.28)$$

where, as before, $\zeta = Z e^{qi}$.

Writing $\xi = X e^{pi}$ and using the general formula

$$g(x) = \frac{1}{2\pi} \int_{-\pi}^{\pi} \frac{\xi g(\xi)}{\xi - x} \, dp,$$

valid for $|\xi| < X$, he obtains

$$F(x, y) + F(x, y_1) + \cdots + F(x, y_{m-1})$$

$$= \frac{1}{4\pi^2} \int_{-\pi}^{\pi} \frac{\xi \, dp}{\xi - x} \int_{-\pi}^{\pi} \zeta \frac{\chi(\xi, b + \zeta)}{f(\xi, b + \zeta)} F(\xi, b + \zeta) \, dq. \qquad (6.29)$$

Writing U_n for the coefficient of x^n in the expansion of the left-hand side of (6.29) in powers of x, he deduces from (6.29) that $|U_n|$ cannot exceed the coefficient of x^n in the expansion of

$$\frac{XZ}{X-x} \wedge \left[\frac{\chi(\xi, b+\zeta)}{f(\xi, b+\zeta)} F(\xi, b+\zeta) \right],$$

the maximum modulus being taken for $|\xi| = X, |\zeta| = Z$.

To obtain explicit formulae for the U_n he returns to (6.28), which gives

$$U_0 = \frac{1}{2\pi} \int_{-\pi}^{\pi} \zeta \frac{\chi(0, b+\zeta)}{f(0, b+\zeta)} F(0, b+\zeta)\, dq,$$

$$U_n = \frac{1}{n!} \frac{1}{2\pi} \int_{-\pi}^{\pi} \zeta \left[D_x^n \left(\frac{\chi(x, b+\zeta)}{f(x, b+\zeta)} \right) F(x, b+\zeta) \right]_{x=0} dq \qquad (n = 1, 2, \ldots).$$

$$(6.30)$$

He transforms these results further by using the identity

$$\frac{\chi(x, b+z)}{f(x, b+z)} = \frac{\chi(0, b+z)}{f(0, b+z)} + D_z \log \frac{f(x, b+z)}{f(0, b+z)},$$

and so arrives at the formula

$$U_n = \frac{1}{n!} \frac{1}{2\pi} \int_{-\pi}^{\pi} \zeta \frac{\chi(0, b+\zeta)}{f(0, b+\zeta)} [D_x^n F(x, b+\zeta)]_{x=0}\, dq$$

$$- \frac{1}{n!} \frac{1}{2\pi} \int_{-\pi}^{\pi} \zeta \left[D_x^n \left(\log \frac{f(x, b+\zeta)}{f(0, b+\zeta)} D_\zeta F(x, b+\zeta) \right) \right]_{x=0} dq,$$

$$(6.31)$$

the second term being obtained by integration by parts.

Cauchy now turns to the important special case where $m = 1$ and the single root of the equation $f(0, b+z) = 0$ is at $z = 0$. For his next argument he uses (without stating it explicitly) the general lemma

$$\frac{1}{2\pi} \int_{-\pi}^{\pi} \zeta h(\zeta)\, dq = \frac{1}{(n-1)!} [D_z^{n-1}(z^n h(z))]_{z=0},$$

which holds if the only singularity of $h(z)$ with $|z| < Z$ is at $z = 0$ and n is at least equal to the multiplicity of any infinity that $h(z)$ has there. After showing that

$$U_0 = F(0, b),$$

he writes down two expressions for U_n $(n = 1, 2, \ldots)$; these are

$$U_n = \frac{1}{(n-1)!n!} \left[D_z^{n-1} \left\{ z^n \left[D_x^n \left(\frac{\chi(b+z)}{f(x, b+z)} F(x, b+z) \right) \right]_{x=0} \right\} \right]_{z=0},$$

$$(6.32)$$

$$U_n = \frac{1}{n!} D_x^n F(x, b) - \frac{1}{(n-1)! n!}$$

$$\times \left[D_z^{n-1} \left\{ z^n \left[D_x^n \left(\log \frac{f(x, b+z)}{f(0, b+z)} D_z F(x, b+z) \right]_{x=0} \right\} \right]_{z=0} \right.$$

(6.33)

In fact, equation (6.33) is correct, but (6.32) (Cauchy's equation (97)) is erroneous,[15] it should read

$$U_n = \frac{1}{(n!)^2} \left[D_z^{n-1} \left\{ z^{n+1} \left[D_x^n \left(\frac{\chi(x, b+z)}{f(x, b+z)} F(x, b+z) \right) \right]_{x=0} \right\} \right]_{z=0} \quad (6.32')$$

The nature of the error can be seen by considering the special case $f(x, b+z) = z - x$, $F(x, b+z) = F(b+z)$. Fortunately, it is equation (6.33) that Cauchy mainly uses for his later results. The erroneous equation also appears in the version in the *Exercices d'analyse* and in the *Oeuvres*.

Cauchy also writes down formulae for the U_n in the more general case where $f(0, b+z)$ has m equal zeros at $z = 0$. He obtains

$$U_0 = mF(0, b)$$

and, denoting by N (which may depend on n) the multiplicity of the infinity of $u_n(z)$ at $z = 0$, he gives expressions for the U_n corresponding to (6.32) and (6.33) above. Here again the first of these (his equation (101)) is erroneous;[16] it should read

$$U_n = \frac{1}{N! n!} \left[D_z^N \left\{ z^N \left[D_x^n \left(\frac{\chi(x, b+z)}{f(x, b+z)} F(x, b+z) \right) \right]_{x=0} \right\} \right]_{z=0}. \quad (6.34)$$

The second one (which is correct) is

$$U_n = \frac{m}{n!} [D_x^n F(x, b)]_{x=0} - \frac{1}{(N-1)! n!}$$

$$\times \left[D_z^{N-1} \left\{ z^N \left[D_x^n \left(\log \frac{f(x, b+z)}{f(0, b+z)} D_z F(x, b+z) \right) \right]_{x=0} \right\} \right]_{z=0}.$$

(6.35)

He then gives a formal theorem[17] (his Theorem V) summarising his results; he concentrates on those contained in equations (6.26), (6.27) and (6.30) above.

---

[15] In Frisiani and Piola's translation of this memoir, they give the correct form (6.32') and, in their commentary, they prove it in detail, but they make no mention of Cauchy's error.

[16] It is corrected by Frisiani and Piola.

[17] [1831d], p. 32; *Exercices d'analyse*, **2**, 76; *Oeuvres* **(2)12**, 87–8.

He concludes by remarking that, if the equation

$$f(0, b + z) = 0$$

has m equal roots at $z = 0$ and $N = n$ for all n, then (6.35) reduces to the formula that was given by Laplace [1779] and corrected by Paoli [1788], but he points out that if $N > n$, then one has to use the general formula (6.35); he gives an example where this is necessary.

6.12.   Cauchy's next result[18] (his Theorem VI) generalises these last results to an annulus in the z-plane. He supposes that the equation

$$f(0, b + z) = 0$$

has m roots in the annulus $z_0 < |z| < Z$, and writes (we have modified his notation slightly)

$$\zeta = Z e^{pi}, \quad \zeta' = z_0 e^{pi}.$$

He also assumes (i) that $f(x, b + z)$ is well defined for $|x| < X$ and $|z| \leq Z$, and (ii) that

$$\frac{\chi(x, b + \zeta)}{f(x, b + \zeta)}, \quad \frac{\chi(x, b + \zeta')}{f(x, b + \zeta')}$$

can be expanded in power-series in x for $|x| < X$. His conclusion is that the equation

$$f(x, b + z) = 0$$

also has m roots in $z_0 < |z| < Z$; if these are $y, y_1, \ldots, y_{m-1}$, then $y + y_1 + \cdots + y_{m-1}$ can be expanded in a convergent series of powers of x. Furthermore, if $F(x, y) = F(x, b + z)$ is 'finite and continuous' when $|x| < X$ and $z_0 < |z| < Z$, then

$$F(x, y) + F(x, y_1) + \cdots + F(x, y_{m-1})$$

can also be expanded in a convergent power-series for $|x| < X$.

Cauchy does not give a full proof of this theorem; he merely indicates that it is based on the formula

$$\int_{-\pi}^{\pi} [\varphi(\xi) - \varphi(\xi')] \, dp = 2\pi \mathscr{E}_{(x_0)}^{(X)} \mathscr{E}_{(-\pi)}^{(\pi)} \left(\left(\frac{\varphi(x)}{x} \right) \right),$$

where $\xi = X e^{pi}$, $\xi' = x_0 e^{pi}$, which comes from his paper [1826g] (Section 5.8 above).

We note that, although Cauchy at this point had in his hands all the tools that would have been needed to prove Laurent's theorem, he did nothing in this direction; we recall that he had given no serious consideration up to this time to isolated essential singularities.

[18] [1831d], p. 35; *Exercices d'analyse*, **2**, 78; *Oeuvres* **(2)12**, 90–1.

6.13. Cauchy now applies his general results to cases where $f(x, y)$ is of the special form

$$f(x, y) = \Pi(y) - x\varpi(y).$$

As we shall see in a moment, he is here coming close to the type of equation that is associated with the Lagrange series.

The conditions of his Theorem V (Section 6.11 above) will be satisfied if $\Pi(b + z)$, $\varpi(b + z)$ and $F(x, b + z)$ are 'finite and continuous' for $|x| < X$ and $|z| \leqslant |Z|$, and

$$\frac{1}{\Pi(b + \zeta) - x\varpi(b + \zeta)}$$

can be expanded in a power-series in x for $|x| < X$; this will certainly hold if

$$X \wedge \frac{\varpi(b + \zeta)}{\Pi(b + \zeta)} \leqslant 1.$$

To get the best value for X, Z has to be chosen so that

$$\wedge \frac{\varpi(b + \zeta)}{\Pi(b + \zeta)}$$

has the smallest possible value;[19] this will be the 'principal modulus', say M, of $\varpi(b + z)/\Pi(b + z)$ (cf. Section 6.9 above). He concludes that if Z is so chosen and $X < 1/M$, then the equations

$$\Pi(b + z) = 0, \tag{6.36}$$

$$\Pi(b + z) - x\varpi(b + z) = 0 \tag{6.37}$$

will have the same number m of roots with $|z| < Z$; if $y, y_1, \ldots, y_{m-1}$ are the values of $y = b + z$ corresponding to the roots of (6.37), then

$$F(x, y) + F(x, y_1) + \cdots + F(x, y_{m-1})$$

can be expanded as a power-series in x for $|x| < X$ and $|z| < Z$; furthermore, the remainder in this series will not exceed

$$\frac{|x|^n Z}{X^{n-1}(X - |x|)} \frac{\wedge \Pi'(b + \zeta) + X \wedge \varpi'(b + \zeta)}{1 - MX} \wedge \frac{F(\xi, b + \zeta)}{\Pi(b + \zeta)}.$$

This result[20] is Cauchy's Theorem VII.

As a corollary, Cauchy shows that if (i)

$$f(y) = \Pi(y) - \varpi(y),$$

(ii) $\Pi(b + z) = 0$ has m roots with $|z| < Z$ and (iii)

$$\wedge \frac{\varpi(b + \zeta)}{\Pi(b + \zeta)} < 1,$$

[19] Cauchy erroneously says the greatest possible value.
[20] [1831d], p. 38; *Exercices d'analyse*, pp. 81–2; *Oeuvres* (2)**12**, 93–4.

then $f(b + z) = 0$ also has exactly m roots with $|z| < Z$. This result is usually known as 'Rouché's theorem' [Rouché 1862].

Cauchy goes on to enumerate some further special cases. He writes out the full results for the case $m = 1$ and, specialising still further to the case

$$\Pi(y) = y - b, \quad F(x, y) = F(y),$$

he obtains the formula

$$F(y) = F(b) + x\varpi(b)F'(b) + \frac{x^2}{2!} D_b\{[\varpi(b)]^2 F'(b)\}$$

$$+ \frac{x^3}{3!} D_b^2\{[\varpi(b)]^3 F'(b)\} + \cdots$$

for the unique solution y with $|y - b| < Z$ of the equation

$$y - b = x\varpi(y);$$

this is precisely the Lagrange series.

He also writes down a formula for the expansion of $F(x, y)$ in this case, but unfortunately in a form (his equation (146)) that we have already seen to be erroneous (cf. Section 6.11 above). Frisiani and Piola give the correct result.

This portion of the memoir essentially concludes with a discussion of the Kepler problem. Cauchy shows that the Lagrange series for the root y of the equation

$$y = b + x \sin y$$

is convergent for $|x| < X$, where

$$X = \frac{Z}{\cosh Z}$$

and Z satisfies the equation

$$Z \tanh Z = 1.$$

It turns out that we have approximately

$$Z = 1.199\,678, \quad X = 0.662\,742;$$

the value of X is close to that given by Laplace [1825], which was 0.661 95 (Section 6.5 above).

The final paragraphs of this portion of [1831d] describe some results, corresponding to those discussed above, for functions of several variables; we shall not try to give an account of these passages, in which no new principles are involved.

6.14. We now turn to [1831e], which we shall usually call Cauchy's second Turin memoir. It originally appeared in lithographed form in Cauchy's handwriting, and it is a photographic reproduction of this that appears in his collected works. The memoir has never been published in printed form in the original French. An Italian translation by Antonio Lombardi appeared in 1839 in the *Memorie di Matematica e di Fisica della Società Italiana delle Scienze residente in Modena*, 22, 91–183 (1839); the translation is not entirely reliable, since Lombardi, who was the Secretary of the Society, was a hydraulic engineer rather than a mathematician, but it does help the reader to decipher some of the less legible passages of the version in Cauchy's collected works. Cauchy published two abstracts [1831b] and [1831c] of this memoir in the *Bulletin de Férussac*.

The memoir is an important one, and it is a pity that it is not more accessible; it is in [1831e] that Cauchy first considers integrals of complex functions round general simple closed curves, and establishes the residue theorem for this case. He also obtains results equivalent to what came to be called the 'principle of the argument'; he proves a more general form of Rouché's theorem, and he devises a 'calculus of indices', in which he considers phenomena that were later to be described in terms of winding numbers.

Cauchy supposes given a function $f(z)$ of the complex variable $z = x + y\mathrm{i}$, tacitly assuming (as usual) that all necessary differentiations can be performed. Regarding (x, y) as rectangular coordinates in the plane, and then going over to general curvilinear coordinates (r, p) he writes down the generalised Cauchy-Riemann equation

$$\frac{\partial}{\partial p}\left[f(z)\frac{\partial z}{\partial r}\right] = \frac{\partial}{\partial r}\left[f(z)\frac{\partial z}{\partial p}\right] = v(p, r),$$

say. This equation was introduced in the 1814 memoir (Section 2.3 above), although the geometrical context did not then appear explicitly.

He now supposes given a closed contour $OO'O''\ldots$ (tacitly assumed to be simple) and considers the repeated integrals

$$\iint v \, \mathrm{d}p \, \mathrm{d}r, \qquad \iint v \, \mathrm{d}r \, \mathrm{d}p, \qquad (6.38)$$

taken over the domain enclosed by the contour. He does not now suppose, as he had in effect done in the 1814 memoir (Section 2.4 above), that the domain is bounded by curves on which either r or p is constant. He argues that, since v is a derivative with respect to p and also with respect to r, each of these repeated integrals will reduce to a single integral, in which the integrand will involve values of $f(z)\partial z/\partial r$ or $f(z)\partial z/\partial p$ only at points on the contour itself. He goes on to say that, if v is finite and continuous for all values of (r, p) in the domain of integration, then the integrals

(6.38) will be equal to each other; in the contrary case, there will be a correction term

$$\Delta = \iint v \, dr \, dp - \iint v \, dp \, dr,$$

whose value can be determined by using the theory of singular integrals or the theory of residues.

He then states the following result: if (i) $x(r, p)$ and $y(r, p)$ are finite and continuous in the domain enclosed by the contour, (ii) the equation

$$1/f(z) = 0$$

has the roots $z_1, z_2, \ldots, z_m$ in this domain and $f(z)$ is otherwise well behaved there, and (iii)

$$\frac{\partial x}{\partial r} \frac{\partial y}{\partial p} - \frac{\partial x}{\partial p} \frac{\partial y}{\partial r} > 0$$

there, then we shall have

$$\Delta = 2\pi i \mathscr{E}((f(z))), \tag{6.39}$$

where $\mathscr{E}((f(z)))$ denotes the sum of the residues of $f(z)$ at $z_1, z_2, \ldots, z_m$.

Cauchy gives no proof for this result, but he does remark that if

$$\frac{\partial x}{\partial r} \frac{\partial y}{\partial p} - \frac{\partial x}{\partial p} \frac{\partial y}{\partial r} < 0$$

at any of the roots $z_1, z_2, \ldots, z_m$, then the sign of the corresponding residue has to be changed, a statement for which he sketches a proof in a footnote. This contains the expression

$$K \int_{-\infty}^{\infty} \left[\frac{1}{\alpha - \rho(\cos \tau + i \sin \tau)} - \frac{1}{\alpha + \rho(\cos \tau + i \sin \tau)} \right] d\alpha,$$

where K is the residue at the root in question. This suggests that he is relying on an argument like that used in Section 4 of [1825b] (Section 4.5 above), where similar expressions occur.

It is curious that Cauchy has returned here to the old methods of the 1814 memoir, with double integrals effectively expressed in curvilinear coordinates. In fact, as we shall see in a moment (Section 6.15 below) he abandons this approach immediately, and works entirely in rectangular coordinates. It is possible that he was remembering his use of polar coordinates in the early part of the first Turin memoir [1831d] (Section 6.9 above), and preparing the ground for further developments in this direction.

6.15. Cauchy's next aim is to convert the correction term Δ into a single integral, which he then uses to obtain an estimate for Δ.

He begins with the case where the contour is a rectangle; he takes $r = x$, $p = y$, and obtains the familiar formula

$$\Delta = - \int_{x_0}^{X} [f(x + Yi) - f(x + y_0 i)] \, dx$$

$$+ i \int_{y_0}^{Y} [f(X + yi) - f(x_0 + yi)] \, dy \tag{6.40}$$

(cf. Section 3.2 above). Putting

$$x = x_0 + (X - x_0)\theta, \quad y = y_0 + (Y - y_0)\theta \quad (0 \leqslant \theta \leqslant 1),$$

he obtains a somewhat messy expression for Δ as an integral with respect to θ over the interval $(0, 1)$. From this he deduces that

$$|\Delta| \leqslant [2(X - x_0) + 2(Y - y_0)]M, \tag{6.41}$$

where M is the maximum of $|f(z)|$ on the perimeter of the rectangle; he notes that the expression in square brackets in (6.41) is precisely the length of the perimeter.

He now turns to the more general case of a simple closed contour. In the expression

$$\int_{x_0}^{X} [f(x + Yi) - f(x + y_0 i)] \, dx$$

appearing in (6.40) above, the constants y_0 and Y now have to be replaced by certain functions of x; if the contour is not convex, one may have to replace the difference

$$f(x + Yi) - f(x + y_0 i)$$

by a sum, which he denotes by V, of such differences. He also denotes arc-length along the contour, measured in the positive direction from a fixed point O, by s and the total perimeter by c; he treats $f(z) = f(x + yi)$ for points (x, y) on the contour as a function of s. The expression

$$\int_{x_0}^{X} [f(x + Yi) - f(x + y_0 i)] \, dx$$

or, in the more general case,

$$\int_{x_0}^{X} V \, dx$$

thus becomes

$$- \int_{0}^{c} f(z) \frac{dx}{ds} \, ds.$$

Similarly, the expression

$$\int_{y_0}^{Y} [f(X + yi) - f(x_0 + yi)] \, dy$$

becomes

$$\int_0^c f(z) \frac{dy}{ds} \, ds.$$

Putting all the terms together, he concludes that

$$\Delta = \int_0^c f(z) \frac{dx + i \, dy}{ds} \, ds$$

$$= \int_0^c f(z) \frac{dz}{ds} \, ds. \qquad (6.42)$$

It now follows from equation (6.39) above that

$$\frac{1}{2\pi i} \int_0^c f(z) \frac{dz}{ds} \, ds = \mathscr{E}((f(z))).$$

Since $|dz/ds| = 1$, it also follows that

$$|\mathscr{E}((f(z)))| \leq \frac{1}{2\pi} \int_0^c |f(z)| \, ds$$

$$\leq \frac{c}{2\pi} \wedge f(z),$$

where $\wedge f(z)$ is the maximum of $|f(z)|$ for values of z lying on the contour.

He then states a formal theorem[21] (his Theorem 1) summing-up these results. He adds that the result still holds if the contour is made up of several portions of straight lines or curves, noting that this would cover cases such as the perimeter of a rectangle, or of a truncated sector specified in polar coordinates (r, p) by $r_0 \leq r \leq R$, $p_0 \leq p \leq P$, where $r_0 < R$ and $-\pi \leq p_0 < P \leq \pi$. He goes on to remark that his results for these cases could be obtained directly from equation (6.41) above; he appears here to be tacitly confessing that his use of general curvilinear coordinates (r, p) (Section 6.14 above) was completely unnecessary.

We see that Cauchy has here established the residue theorem for a general simple closed contour. We note also that his arguments follow the general pattern of those that are commonly used to prove the divergence theorem, expressing a double integral over a domain in terms of a line integral round the perimeter of the domain.

---

[21] [1831e], p. 10; *Oeuvres* **(2)15**, 191.

6.16. For his main application of the general residue theorem, Cauchy assumes that the functions $f_0(z)$, $f_0'(z)$ and $F(z)$ are 'finite and continuous' inside and on the closed contour $OO'O''\ldots$ and writes $z_1, z_2, \ldots, z_m$ for the roots of the equation $f_0(z) = 0$ inside the contour. He concludes that

$$F(z_1) + F(z_2) + \cdots + F(z_m) = \mathscr{E}\left(\left(\frac{f_0'(z)}{f_0(z)}F(z)\right)\right)$$

$$= \frac{1}{2\pi i}\int_0^c \frac{f_0'(z)}{f_0(z)}F(z)\frac{dz}{ds}\,ds.$$

In particular, we shall have

$$z_1 + z_2 + \cdots + z_m = \frac{1}{2\pi i}\int_0^c z\frac{f_0'(z)}{f_0(z)}\frac{dz}{ds}\,ds$$

and

$$m = \frac{1}{2\pi i}\int_0^c \frac{f_0'(z)}{f_0(z)}\frac{dz}{ds}\,ds.$$

It also follows that

$$|F(z_1) + F(z_2) + \cdots + F(z_m)| \leqslant \frac{c}{2\pi}\wedge\left(\frac{f_0'(z)}{f_0(z)}F(z)\right),$$

the upper bound $\wedge$ being taken over the values of z lying on the contour. He writes down the corresponding results for $|z_1 + z_2 + \cdots + z_m|$ and for m, and states all these results as a formal theorem.

We note that Cauchy here requires $f_0(z)$ to have a continuous derivative, but he does not seem to attribute any special significance to the introduction of this condition.

6.17. Cauchy now begins a detailed analysis of the formula

$$m = \frac{1}{2\pi i}\int_0^c \frac{f_0'(z)}{f_0(z)}\frac{dz}{ds}\,ds. \tag{6.43}$$

For values of z lying on the contour, he expresses $f_0(z)$ as a function of s, say

$$f_0(z) = \varphi(s) + i\chi(s),$$

so that (6.43) becomes

$$m = \frac{1}{2\pi i}\int_0^c \frac{\varphi'(s) + i\chi'(s)}{\varphi(s) + i\chi(s)}\,ds.$$

He supposes first that $\varphi(s)$ is of constant sign on an interval $s_0 \leqslant s \leqslant s_1$; if $\varphi(s) > 0$ there, then

$$\int_{s_0}^{s_1} \frac{\varphi'(s) + i\chi'(s)}{\varphi(s) + i\chi(s)}\,ds = \log\left[\varphi(s_1) + i\chi(s_1)\right] - \log\left[\varphi(s_0) + i\chi(s_0)\right],$$

the logarithms being given their principal value; on the other hand, if $\varphi(s) < 0$ in the interval, then

$$\int_{s_0}^{s_1} \frac{\varphi'(s) + i\chi'(s)}{\varphi(s) + i\chi(s)} \, ds = \log\left[-\varphi(s_1) - i\chi(s_1)\right] - \log\left[-\varphi(s_0) - i\chi(s_0)\right].$$

He combines the two cases in the single formula[22]

$$\int_{s_0}^{s_1} \frac{\varphi'(s) + i\chi'(s)}{\varphi(s) + i\chi(s)} \, ds = \log\left[\frac{\varphi(s_1)}{\varphi(s_0)}\right] + \log\left[1 + \frac{i\chi(s_1)}{\varphi(s_1)}\right] - \log\left[1 + \frac{i\chi(s_0)}{\varphi(s_0)}\right].$$

Since $\varphi(0) = \varphi(c)$ and $\chi(0) = \chi(c)$, it follows at once that if $\varphi(s)$ is of constant sign on the whole contour, then $m = 0$.

To find out what happens when $\varphi(s)$ suffers changes of sign as s varies on the contour, Cauchy supposes that the interval (s_0, s_1) contains a single root σ of the equation $\varphi(s) = 0$. Supposing first that $\varphi(s)$ changes from negative to positive values as s increases through σ, he writes down the approximate equation

$$\int_{s_0}^{s_1} \frac{\varphi'(s) + i\chi'(s)}{\varphi(s) + i\chi(s)} \, ds = \int_{s_0}^{\sigma - \varepsilon} + \int_{\sigma + \varepsilon}^{s_1}$$

$$= \log\left[\varphi(s_1) + i\chi(s_1)\right] - \log\left[-\varphi(s_0) - i\chi(s_0)\right]$$

$$- \log\left[\varphi(\sigma + \varepsilon) + i\chi(\sigma + \varepsilon)\right]$$

$$+ \log\left[-\varphi(\sigma - \varepsilon) - i\chi(\sigma - \varepsilon)\right],$$

where ε is small and positive. The terms involving ε give

$$\log\left[-\varphi(\sigma - \varepsilon) - i\chi(\sigma - \varepsilon)\right] - \log\left[\varphi(\sigma + \varepsilon) + i\chi(\sigma + \varepsilon)\right] = E,$$

say. Provided that $\chi(\sigma) \neq 0$, this reduces to

$$E = \log\left[1 + \frac{i\chi(\sigma - \varepsilon)}{\varphi(\sigma - \varepsilon)}\right] - \log\left[1 + \frac{i\chi(\sigma + \varepsilon)}{\varphi(\sigma + \varepsilon)}\right]. \tag{6.44}$$

Equation (6.44) holds also when $\varphi(s)$ changes from positive to negative as s increases through σ. After examining the case $\chi(\sigma) = 0$, Cauchy concludes that

$$E = \mp\pi i$$

according as the ratio $\chi(s)/\varphi(s)$ changes from negative to positive or from positive to negative as s increases through σ; if the ratio does not change sign, then $E = 0$.

In the general situation, if $\varphi(s)$ vanishes for several values of s in the interval (s_0, s_1), then it follows that

---

[22] There are some serious misprints in this formula, both in the original lithograph and in the Italian translation; we have corrected them here.

$$\int_{s_0}^{s_1} \frac{\varphi'(s) + i\chi'(s)}{\varphi(s) + i\chi(s)}\, ds = \log\left|\frac{\varphi(s_1)}{\varphi(s_0)}\right| + \log\left[1 + \frac{i\chi(s_1)}{\varphi(s_1)}\right]$$

$$-\log\left[1 + \frac{i\chi(s_0)}{\varphi(s_0)}\right] + \pi i(m'' - m'),$$

where m' and m'' are, respectively, the number of zeros of $\varphi(s)$ in (s_0, s_1) at which $\chi(s)/\varphi(s)$ changes from negative to positive and the number at which it changes from positive to negative.

In particular, integration round the whole contour gives

$$\int_0^c \frac{\varphi'(s) + i\chi'(s)}{\varphi(s) + i\chi(s)}\, ds = \pi i(m'' - m'),$$

whence, finally, the number of zeros of $f_0(z)$ inside the contour is given by

$$m = \tfrac{1}{2}(m'' - m').$$

6.18. The results described in the previous section lead Cauchy to set up the following formal definition.

Let $\psi(s)$ be a real-valued function of a real variable s (no longer necessarily denoting arc-length on a closed contour); if σ is a root of the equation

$$1/\psi(s) = 0,$$

then the *index* of ψ with respect to σ is

$$\tfrac{1}{2}[\operatorname{sgn} \psi(\sigma + \varepsilon) - \operatorname{sgn} \psi(\sigma - \varepsilon)],$$

where ε is small and positive. Here we have written $\operatorname{sgn} u = \pm 1$ according as $u \gtrless 0$, and $\operatorname{sgn} 0 = 0$; Cauchy writes $u/\sqrt{(u^2)}$ instead of our $\operatorname{sgn} u$. He then defines the *integral index* of $\psi(s)$ over the interval (s_0, s_1) to be the sum of the indices of $\psi(s)$ with respect to the zeros of $1/\psi(s)$ in (s_0, s_1), and denotes it by

$$\mathcal{I}_{s_0}^{s_1}((\psi(s))).$$

Using this notation, he expresses his main result in the following form[23] (his Theorem 3).

Let $OO'O'' \ldots$ be a (simple) closed contour in the (x, y)-plane, and suppose that $f_0(z)$ and $f_0'(z)$ are finite and continuous for values of $z = x + yi$ corresponding to points inside or on the contour. Let $f_0(z)$ be expressed as a function of the arc-length s, say

$$f_0(z) = \varphi(s) + i\chi(s),$$

[23] [1831e], p. 22; *Oeuvres* (2)**15**, 203.

for points on the contour; then the number m of zeros of $f_0(z)$ inside the contour is given by

$$m = -\tfrac{1}{2}\mathscr{I}_0^c\left(\left(\frac{\chi(s)}{\varphi(s)}\right)\right).$$

We see that Cauchy is using the index to perform functions similar to those for which the winding number is used in later treatments of the situation.

He gives some formal properties of the notion of index. He remarks that if

$$f_0(z) = f_1(z) \cdot f_2(z) \ldots,$$

then

$$\frac{f_0'(z)}{f_0(z)} = \frac{f_1'(z)}{f_1(z)} + \frac{f_2'(z)}{f_2(z)} + \cdots;$$

hence, if $f_0(z)$, $f_1(z)$, $f_2(z)$, $\ldots$ have, respectively, m, m_1, m_2, $\ldots$ zeros inside the contour, then

$$m = m_1 + m_2 + \cdots,$$

where we have

$$m_1 = \frac{1}{2\pi i}\int_0^c \frac{f_1'(z)}{f_1(z)}\frac{dz}{ds}\,ds,$$

and so on.

In particular, if we take

$$f_1(z) = \frac{f_0(z)}{\cos\tau + i\sin\tau}, \qquad f_2(z) = \cos\tau + i\sin\tau,$$

where τ is a constant angle, then $m_1 = m$ and $m_2 = 0$. On the contour, we have

$$f_1(z) = \varphi(s)\cos\tau + \chi(s)\sin\tau + i[\chi(s)\cos\tau - \varphi(s)\sin\tau],$$

whence

$$m = m_1 = \tfrac{1}{2}\mathscr{I}_0^c\left(\left(\frac{\varphi(s)\sin\tau - \chi(s)\cos\tau}{\varphi(s)\cos\tau + \chi(s)\sin\tau}\right)\right).$$

The special case $\tau = \pi/2$ then gives the alternative simple formula

$$m = \tfrac{1}{2}\mathscr{I}_0^c\left(\left(\frac{\varphi(s)}{\chi(s)}\right)\right).$$

6.19. Cauchy now proceeds to obtain some applications of this result. He begins by proving the following theorem[24] (his Theorem 4). If

[24] [1831e], p. 24; *Oeuvres* (2)15, 205.

$$f_0(z) = \Pi(z) + \varpi(z),$$

where $f_0(z)$, $\Pi(z)$ and $\varpi(z)$ are finite and continuous inside and on the closed contour $OO'O''\ldots$, together with their derivatives, and $\wedge(\varpi(z)/\Pi(z)) < 1$, the upper bound being taken over the values of z on the contour, then the two equations $f_0(z) = 0$ and $\Pi(z) = 0$ have the same number of roots inside the contour. This result is usually known as Rouché's theorem, although Rouché only gave it for the case where the contour is a circle (Rouché [1862]).

To prove this, he writes $f_0(z) = f_1(z)f_2(z)$, where

$$f_1(z) = \Pi(z), \quad f_2(z) = 1 + \frac{\varpi(z)}{\Pi(z)};$$

since the real part of $f_2(z)$ is positive when z is on the contour, it follows that the number m_2 of its zeros inside the contour is 0, and we have

$$m = m_1 + m_2 = m_1,$$

as required.

Having established this result, Cauchy goes on to obtain explicit formulae involving the zeros of $f_0(z)$ and $\Pi(z)$. He denotes the zeros of $f_0(z)$ by $z_1, z_2, \ldots, z_m$ and those of $\Pi(z)$ by $\zeta_1, \zeta_2, \ldots, \zeta_m$, and supposes that a given function $F(z)$ is 'finite and continuous' inside and on the contour. His aim is to relate the expressions

$$F(z_1) + F(z_2) + \cdots + F(z_m), \quad F(\zeta_1) + F(\zeta_2) + \cdots + F(\zeta_m).$$

By an earlier result (Section 6.16 above), we have

$$F(z_1) + F(z_2) + \cdots + F(z_m) - F(\zeta_1) - F(\zeta_2) - \cdots - F(\zeta_m)$$

$$= \mathscr{E}\left(\left(F(z)\frac{f_0'(z)}{f_0(z)}\right)\right) - \mathscr{E}\left(\left(F(z)\frac{\Pi'(z)}{\Pi(z)}\right)\right)$$

$$= \mathscr{E}\left(\left(F(z)\frac{\mathrm{d}}{\mathrm{d}z}\log\left(1 + \frac{\varpi(z)}{\Pi(z)}\right)\right)\right). \tag{6.45}$$

Since $\wedge(\varpi(z)/\Pi(z)) < 1$, we can expand

$$\log\left(1 + \frac{\varpi(z)}{\Pi(z)}\right),$$

for values of z on the contour, in the convergent series

$$\sum_{n=1}^{\infty} \frac{(-1)^{n-1}}{n}\left(\frac{\varpi(z)}{\Pi(z)}\right)^n.$$

Regarding the integral residue on the right-hand side of (6.45) as equivalent to an integral round the contour, we can integrate term by term, obtaining

$$F(z_1) + \cdots + F(z_m) - F(\zeta_1) - \cdots - F(\zeta_m)$$

$$= \sum_{n=1}^{\infty} \frac{(-1)^{n-1}}{n} \mathscr{E}\left(\left(F(z)\frac{d}{dz}\left(\frac{\varpi(z)}{\Pi(z)}\right)^n\right)\right)$$

$$= \sum_{n=1}^{\infty} \frac{(-1)^n}{n} \mathscr{E}\left(\left(F'(z)\left(\frac{\varpi(z)}{\Pi(z)}\right)^n\right)\right), \tag{6.46}$$

by an earlier result about the residue of a product of functions [1826e] (Section 5.7 above). The result could also be obtained directly by integration by parts.

Writing u_n for the nth term of the series (6.46), and also

$$\wedge \frac{\varpi(z)}{\Pi(z)} = M < 1, \quad \wedge F'(z) = N,$$

Cauchy deduces that

$$|u_n| \leqslant \frac{cN}{2\pi} \frac{M^n}{n}, \tag{6.47}$$

where c is the length of the contour; he then uses (6.47) to obtain an estimate for the remainder after n terms in the series (6.46). He also notes the equation

$$z_1 + z_2 + \cdots + z_m = \zeta_1 + \zeta_2 + \cdots + \zeta_m$$

$$+ \sum_{n=1}^{\infty} \frac{(-1)^n}{n} \mathscr{E}\left(\left(\left(\frac{\varpi(z)}{\Pi(z)}\right)^n\right)\right).$$

This group of results forms[25] his Theorem 5.

He then examines the special case where

$$\Pi(z) = (z - a)^m.$$

Equation (6.46) thus becomes

$$F(z_1) + \cdots + F(z_m)$$

$$= mF(a) + \sum_{n=1}^{\infty} \frac{(-1)^n}{n} \mathscr{E}\left(\left(F'(z)\frac{(\varpi(z))^n}{(z-a)^{mn}}\right)\right)$$

$$= mF(a) + \sum_{n=1}^{\infty} \frac{(-1)^n}{n} \frac{1}{(mn-1)!} \frac{d^{mn-1}}{da^{mn-1}} \{[\varpi(a)]^n F'(a)\},$$

provided that the condition

[25] [1831e], pp. 27–28; *Oeuvres* (2)15, 208–9.

$$\wedge\frac{\varpi(z)}{(z-a)^m}<1$$

is satisfied. Cauchy had previously given this result [1824], without proof, for the case where $\varpi(z)$ and $F(z)$ are polynomials, and the series collapses to a finite sum.

He finally notes that if $m = 1$ and

$$\wedge\frac{\varpi(z)}{z-a}<1,$$

so that the equation

$$z-a+\varpi(z)=0$$

has a unique root z inside the contour, then we have

$$F(z) = F(a) + \sum_{n=1}^{\infty}\frac{(-1)^n}{n!}\frac{\mathrm{d}^{n-1}}{\mathrm{d}a^{n-1}}[(\varpi(a))^n F'(a)],$$

which is precisely the Lagrange series once more.

6.20. We need not examine the remainder of [1831e] in detail, since it does not contribute anything to the development of the general theory of complex functions. There are detailed descriptions of methods for evaluating upper bounds $\wedge g(z)$ and indices $\mathscr{I}((\psi(s)))$ in various special cases; there are then some applications of Cauchy's index theory to problems about the real zeros of real polynomials, showing how known results by Descartes, Budan, Fourier and Sturm can be established in this way. Cauchy finally returns to the expansions we discussed in Section 6.19 above, giving numerical examples of their use for the calculation of roots of equations and for error estimates.

Cauchy's calculus of indices reappears in the memoir [1833a]; here he treated the topic without any reference to complex integration. A much extended version was published in the *Journal de l'Ecole Polytechnique* (volume 15, cahier 25, pp. 176–229) in 1837, and there is an Italian translation of the original lithographed version in *Mem. Soc. Ital. Sci. Modena* **(1)22**, 228–46 (1839).

6.21. We now summarise what was achieved in the papers we have described in the present chapter.

The Lagrange series was introduced by Lagrange in 1770 in the context of the solution of algebraic equations. Supposing that p is a root of an equation of the form

$$a-x+\varphi(x)=0,$$

where $\varphi(x)$ is a polynomial, and that $\psi(x)$ is a given function, he obtained the formal expansion

$$\psi(p) = \psi(\alpha) + \sum_{n=1}^{\infty} \frac{1}{n!} \frac{d^{n-1}}{d\alpha^{n-1}} [\varphi^n(\alpha)\psi'(\alpha)].$$

In 1771 he used it for transcendental functions, with the aim of obtaining a solution of Kepler's problem in celestial mechanics, i.e. the problem of finding an expression for $\psi(u)$ when u satisfies the equation

$$t = u - c \sin u.$$

The convergence of the series was first studied, in a particular case, by Laplace in 1825. Cauchy than embarked on a thorough investigation of the problem. In 1827 he obtained an asymptotic expansion for the terms of the Lagrange series in a general context, and in the course of this work he used the expressions that he had already discovered for the higher derivatives of analytic functions in terms of contour integrals taken round circles. It was probably at this stage that Cauchy realised that he was in possession of a remarkably powerful tool.

Cauchy exploited this tool in a major fashion in his first Turin memoir in 1831. He began by proving that the Maclaurin series of an analytic function is convergent up to the singularity nearest to the origin, also obtaining estimates for the terms of the series and for its remainder after a finite number of terms. In particular, he proved what was to become known as "Cauchy's inequality", i.e.

$$\frac{|f^{(n)}(0)|}{n!} \leq \frac{\max |f(\zeta)|}{R^n},$$

the maximum being taken over the circle $|\zeta| = R$, which holds provided that R is smaller than the modulus of the singularity of f nearest to the origin. This result became the basis for what Cauchy called the 'calcul des limites', better known later as the method of majorants.

He then went on to obtain results on the expansion of implicit functions. On the hypothesis that an equation $f(x, y) = 0$, where f is an analytic function, has m roots $y, y_1, \ldots, y_{m-1}$ inside a given circle, he obtains convergent power series to represent expressions of the form

$$F(y) + F(y_1) + \cdots + F(y_{m-1}).$$

The Lagrange series then appears as a particular case when one supposes that $m = 1$.

In the second Turin memoir of 1831, he studies general simple closed contours for the first time, proving the residue theorem for this situation. Here again he finds expressions for sums such as

$$F(z_1) + F(z_2) + \cdots + F(z_m),$$

where $z_1, z_2, \ldots, z_m$ are the zeros of a function $f(z)$ inside a given contour. In particular, he establishes formulae for the number of zeros; in discussing these he creates what he calls a 'calcul des indices', involving concepts closely related to later ideas about winding numbers. Among the results he obtains in this context is the general form of Rouché's theorem, which he applies to prove the convergence of the Lagrange series in some important special cases.

It is curious that Cauchy never republished the second Turin memoir in printed form. It may be that he regarded his later reformulation of the 'calcul des indices' without bringing his theory of residues into the picture as being more important. It is clear that in this whole group of papers his main aim is to obtain results about the solutions of algebraic or transcendental equations, and that he regards the theorems he has proved in general complex function theory as being in a sense peripheral to this principal object.

7

Summary and conclusions

7.1. In the last six chapters we have analysed Cauchy's work on complex function theory from 1814 to 1831, and we have indicated some of the background to the development of his ideas. In the present chapter we shall try to draw the threads together. There are several themes that recur throughout the story; we shall look at these in turn, and try to trace how Cauchy's approach to each of them changed during the period.

To begin with, we shall have to say something about the kind of functions that Cauchy regarded as admissible, and then about the development of his concept of the integral. After a brief sketch of some earlier ideas about complex functions and about techniques for the evaluation of definite integrals, we shall show how, by combining some of these ideas, he was able to obtain results equivalent to special cases of 'Cauchy's theorem' and the residue theorem. We shall then follow the development of his techniques, which enabled him to establish more general forms of these results, and we shall indicate the role played here by his tentative introduction and gradually increasing use of a geometrical picture of the complex plane.

We shall describe some of the applications he made of his results, initially to the evaluation of definite integrals, and then to the summation of infinite series and to the discovery of explicit expressions for the roots of equations; we shall also show how the methods he used here led him to discover his major theorem on the convergence of Taylor series, together with his results on the Lagrange series and series for the representation of implicit functions.

We shall also examine some peripheral issues that crop up persistently, and we shall draw attention to some tentative anticipations of later results on the representation of analytic functions.

7.2. We recall that in the eighteenth century a function was usually thought of as being defined by an analytic expression; by the principle of the generality of analysis, which was widely and often tacitly accepted, such an expression was expected to be valid for all values, real or complex, of the independent variable.

Cauchy makes some general remarks about functions in Chapter I, Section 1, of his *Analyse algébrique* [1821a]. From what he says there, it is clear that he normally regards a function as being defined by an analytic expression (if it is explicit) or by an equation or a system of equations (if it is implicit); where he differs from his predecessors is that he is prepared to consider the possibility that a function may be defined only for a restricted range of values of the independent variable. In particular, he rejects the notion, based on the principle of the generality of analysis, that a function that is initially defined for real values of the variable has an automatic extension to complex values; in Chapter VIII of the *Analyse algébrique* he is careful to provide explicit definitions for the elementary functions of a complex variable.

Cauchy's formal definition of the notion of a continuous function appears in Chapter II, Section 2, of the *Analyse algébrique*, but the idea is clearly present in embryonic form much earlier; in Part II, Section III, of his 1814 memoir on definite integrals he speaks of a function 'increasing or decreasing in a continuous manner' (cf. Section 2.9 above).

In his earlier papers Cauchy makes no explicit hypotheses about the functions he considers, beyond saying that they should be 'determinate'; in practice, he tacitly assumes that they are sufficiently well behaved for them to be differentiated as often as may be required. From the publication of his *Calcul infinitésimal* [1823b] onwards, his usual hypothesis on a function is that it should be 'finite and continuous' (see, for example, Sections 3.6 and 4.4 above), but he still takes for granted that it will normally have continuous derivatives of any required order; he does so even when he is dealing with functions of a complex variable, a clear indication that he is thinking in terms of analytic expressions.

Although some of his early results, even in the 1814 memoir (Section 2.16 above) suggest that in certain circumstances one might consider a function given by different expressions in different intervals, Cauchy seems reluctant to introduce such a function deliberately; even when, in the 1825 memoir, he considers a path in the complex plane defined by a succession of different expressions or equations, he goes to considerable trouble to avoid saying that the path could be specified by a single function (Section

4.12 above). In a postscript added at the end of [1828], however, he extends the definition of a function given in a certain interval by requiring it to vanish outside the interval, and quotes Fourier as his source for the device.

He first contemplates the possibility that a continous function may fail to have a continuous derivative in [1839], remarking that one cannot be certain that a function and its derivative necessarily have discontinuities or infinities for the same values of the independent variable (Section 6.9 above). Here he may perhaps have been influenced by some comments made by Frisiani and Piola in the appendix to their Italian translation (1834) of the first Turin memoir [1831d]. It is clear that he was concerned only by the possibility that the derivative might have discontinuities at isolated points. In any case, he did insert the additional hypothesis that the derivative should be continuous when [1831d] was republished in his *Exercices d'analyse* in 1841, although it appears that he did not do so consistently in later publications [Belhoste 1991, p. 204].

It was not until about 1851 that Cauchy came to recognise $u(x, y) + iv(x, y)$, where u and v are general continuous functions of (x, y), as being a continuous function of $z = x + iy$ [Belhoste 1991, p. 234]; it was only then that he realised the significance of the relation between the Cauchy–Riemann equations and the differentiability of a function of a complex variable. Riemann's dissertation, in which he explored the same complex of ideas, dates from that same year 1851.

7.3. We recall that, for most of the eighteenth century, the indefinite integral was identified with the primitive function and the definite integral with the difference between two values of the primitive function. The fact that

$$\int_a^b f(x)\,\mathrm{d}x$$

was equal in some sense to the sum of the differential elements $f(x)\,\mathrm{d}x$, which was approximately its original definition, came to be regarded as a theorem that required proof.

Some paradoxical results had been noticed; d'Alembert, Lacroix and Lagrange drew attention to cases where a positive function having an infinity in the range of integration appeared to have a negative definite integral, or a real function an imaginary definite integral. As Lagrange remarked, the principles of the differential calculus failed to be applicable (Section 2.9 above).

In his 1814 memoir Cauchy examined some cases of this kind. He suggested that, if necessary, the integral should be redefined to make it a continuous function of its upper limit, and devised a correction term to bring this about (Section 2.9 above). His technique here foreshadows his formal definition in 1822 of the principal value of a definite integral (Section 3.8 above); in fact, several principal-value integrals had already been evaluated in the 1814 memoir (Section 2.14 above). In his 1822 paper he insists that a definite integral should always be regarded as a sum of differentials; in support of this thesis he argues that such a definition can be used even when no primitive function is known, and that it will always give real values for the integrals of real functions (Section 3.8 above).

Finally, in his *Calcul infinitésimal* in 1823 he gives his formal definition of the definite integral as a limit of approximating sums, and deduces its elementary properties; a little later he extends his definition to cover the integration of complex functions of a real variable (Section 3.3 above). From then on he was to use this definition consistently throughout his work.

7.4. In the course of the eighteenth century there was growing interest in the evaluation of definite integrals, especially over an infinite range, in cases where no primitive function was available.

Three main techniques were used. These were (i) changing the order of integration in a double integral (Section 1.12 above), (ii) obtaining and solving a differential equation with respect to a parameter appearing in the integrand (Section 1.12 above) and (iii) the use of 'imaginary substitutions'; in other words, making a complex change of variable. This third method was pioneered by Euler in 1781 (Section 1.9 above) and was used by Laplace in 1782, independently of Euler, in his work on approximations to functions of large numbers (Section 1.10 above); it was later to be interpreted in terms of integration along paths in the complex plane. Laplace returned to this technique from 1809 onwards, making considerable use of it in his work on probability theory. This triggered a lively exchange of views between Poisson and Laplace, which continued for several years. Poisson maintained that this method could at best be regarded as a kind of induction that might be useful for discovering new results, and insisted that results obtained by it should be confirmed in a more direct way; Laplace defended the method by appealing to the principle of the generality of analysis, but was eventually constrained to admit that direct confirmation was desirable (Sections 1.10–1.16 above). This controversy formed the immediate background of Cauchy's first memoir on definite integrals in 1814.

Another important eighteenth-century contribution to complex function theory was the discovery of the 'Cauchy–Riemann equations'. They first appeared in 1752 in d'Alembert's work on plane fluid flow (Section 1.4 above), and reappeared several times in the same context in papers by Euler, d'Alembert and Lagrange (Section 1.5 above); it was recognised that solutions of them could be constructed by considering explicit functions of a complex variable. Similar but more general equations appeared in a 1768 paper by Euler on orthogonal trajectories and in his 1775 paper on the conformal mapping of a sphere on a plane (Section 1.6 above). Euler also used them in a series of papers written in 1777 to show that certain complicated functions of a real variable could be integrated in finite terms (Section 1.8 above).

7.5. In this section we shall trace the gradual development of the ideas and techniques that led Cauchy to the general form of the results that became known as 'Cauchy's theorem' and the residue theorem.

It could be argued that, from the point of view that was current before Cauchy's time, Cauchy's theorem, expressed as the statement that the definite integral of a function between limits is uniquely determined, is almost a triviality, for such an integral was regarded as being the difference between the values of a primitive function at the two limits. This idea is implicit in Euler's 1777 papers on complex integration (Section 1.8 above), although he actually considers only indefinite integrals. The only caveat that might have been raised to this statement would have arisen from the example given by d'Alembert in 1768 (Section 1.3 above).

In Section 7.4 above we have drawn attention to the controversy between Poisson and Laplace about the validity of the method of imaginary substitutions; it is not unlikely that Laplace may have suggested to Cauchy that he should look into the problem. Cauchy's long memoir on definite integrals was completed in 1814, but was not published till 1827; we shall usually refer to it as the 1814 memoir. In his introduction to it Cauchy described his aim as the justification of results obtained by imaginary substitutions by a 'direct and rigorous analysis' (Section 2.2 above).

The results that Cauchy obtained in Part I of the 1814 memoir can be seen to be equivalent to 'Cauchy's theorem' for a class of curvilinear quadrilaterals (Sections 2.3 and 2.4 above). His main tools were (i) a generalised form of the 'Cauchy–Riemann equations' and (ii) changing the order of integration in a double integral; as we have seen in Section 7.4, both these ideas had appeared before. The significance of these results was somewhat obscured by his insistence on expressing them throughout in terms of integrals of real-valued functions over real intervals, possibly with the aim of demonstrating how little use he needed to make of imaginary expressions. A side-effect of this approach was that he ran into complications in dealing with integrals involving trigonometric functions; to cope with these he invented a special technique, which he called the 'separation of exponentials' (Section 2.7 above).

He entered new territory in Part II of the 1814 memoir. He found that when the function with which he was dealing has an infinity (later called a pole) in the domain under consideration, the repeated integrals used in Part I may fail to be equal (Section 2.10 above). He therefore introduced a correction term to allow for the difference between them; in the case of a simple pole, he showed that this term can be expressed as what he calls a 'singular integral', which can be thought of as a non-zero definite integral takes over an infinitely small interval (Section 2.11 above). His results can be seen to be equivalent to the residue theorem for functions having only simple poles and for the class of quadrilaterals arising from Part I (Section 2.12 above); his correction terms are, in fact, equal (apart from a constant factor 2π) to the real and imaginary parts of the sum of the residues at the relevant poles of the function.

It is clear from his expression of surprise (Section 2.12 above) that the correction terms depend only on the function considered and its singularities, but not on the transformation of variables (which carried a rectangle into a curvilinear quadrilateral), that at this stage he had not achieved a full understanding of what was going on; indeed, his values for the correction terms emerge only at the end of some complicated analytic manipulation.

In the 1814 memoir, most of the applications that he gives for his results are to the evaluation of definite integrals, in many cases over an infinite interval.

Between 1814 and 1824 Cauchy made a number of refinements in the methods of the 1814 memoir. Probably by 1817, and almost certainly by 1819, he was using integrals of complex-valued functions over real intervals, abandoning his former restriction to real integrands (Section 3.3 above); in 1823 he gave a formal definition for the integral of a complex function over a real interval. As we can see from the footnotes added to the 1814 memoir before its ultimate publication in 1827, this step resulted in substantial simplifications in his exposition (Section 3.2 above). He emerged with a single correction term, which we can recognise as being (apart from a factor $2\pi i$) the sum of the residues at the poles in the domain; in particular, he was able to dispense with his special technique of the separation of exponentials.

He restated his results several times in 1822–23 (Section 3.5 above). In the *Calcul infinitésimal* in 1823 he gave a somewhat simpler (but still messy) derivation of the correction term for the case of a rectangle; in his proof he replaced the 'Cauchy–Riemann equations' by the single equation

$$i \frac{\partial}{\partial x} f(x + yi) = \frac{\partial}{\partial y} f(x + yi).$$

He gave some attention in 1822–23 to the evaluation of integrals involving polar coordinates (Section 3.7 above). Some of his results, such as the equation

$$\int_{-\pi}^{\pi} \frac{f(e^{ip}) \, dp}{1 - a e^{ip}} = 2\pi[f(0) - f(1/a)] \qquad (a^2 > 1),$$

look as if he is integrating round a circle, but it is clear from the context that he is not thinking in such terms. Poisson had obtained similar formulae a little earlier by different methods.

The correction term (essentially the residue) associated with a multiple pole appears for the first time in an 1823 paper, but without proof (Section 3.10 above).

The next major advance in Cauchy's development of complex function theory comes in his famous 1825 memoir on integration between imaginary limits. In this he defines the integral of a function along a path joining two complex numbers as a limit of approximating sums. Initially he requires the real and imaginary parts of the independent variable to vary monotonically along the path (Section 4.3 above), but this condition is substantially weakened at a later stage (Section 4.12 above) and does not reappear in his later work.

As we have already mentioned in Section 7.4, much of the early work on the evaluation of definite integrals by imaginary substitutions can be interpreted in term of integration along paths in the complex plane. In 1820, in an attempt to reconcile the two conceptions of the definite integral, Poisson considered the possibility of allowing the independent variable to proceed between the two limits of integration through a succession of imaginary values (Section 4.2 above); he also drew attention to cases where the value of an integral is changed by making an imaginary substitution. This paper of Poisson's, together with some unpublished work by Brisson, appears to have stimulated Cauchy to make a serious study of integrals between imaginary limits. An 1822 reference by Cauchy to definite integrals 'between real limits' suggests that his thoughts were already moving in this direction (Section 3.8 above).

The first main result in the 1825 memoir is that the integrals of a function along two neighbouring paths joining the same limits are equal, provided that the function is 'finite and continuous'; his proof is a kind of infinitesimal homotopy argument, which he then rephrases in the language of the calculus of variations (Section 4.4 above).

He then tackles the problem of finding the difference between the integrals along two neighbouring paths when the integrand has an infinity between the paths; in the case of a simple pole, he allows it to lie on one of the paths, so that the integral has to be interpreted as a principal value. He gives two proofs of his results for the case of a simple pole, and then two proofs for a multiple pole (Sections 4.5 and 4.6 above). In each case one of his proofs is modelled on the one he gave for a simple pole in the *Calcul infinitésimal* in 1823. The alternative proofs introduce the new device of subtracting from the function a simple rational expression for its singular part, and dealing with that separately; there is some evidence that this device may have been suggested to him by M. V. Ostrogradskiĭ.

In the middle of the 1825 memoir Cauchy suddenly introduces some geometrical language, representing the complex number $x + yi$ by the point in the plane with Cartesian coordinates (x, y) (Section 4.8 above). Up to this time he had (following the example of Lagrange) scrupulously avoided any use of geometrical language in the context of analysis, believing, it seems, that its use would encourage appeals to intuition and tend to disguise the general applicability of analytic results. At this point, however, he seems to find that he can express some of his ideas more concisely by admitting geometrical terminology. He does not, on the other hand, mention the familiar geometrical interpretations for the addition and multiplication of complex numbers. What he does is to use some geometrical ideas to discuss what happens when two paths cross each other, and again to clarify a homotopy argument in which he compares integrals along two paths that are not close neighbours. At first he uses the language sparingly, but from 1827 onwards he no longer hesitates to bring in geometrical concepts (Section 5.10 above).

Most of the applications that he makes of the results of the 1825 memoir are to the evaluation of definite integrals and the summation of infinite series (Section 4.10 above).

In 1826 Cauchy began to publish his *Exercices de mathématiques*; he developed his general theory further in several of the papers that appeared there. He begins by defining the notion of the residue of a function at an infinity, i.e. a pole (Section 5.2 above); he introduces some special notation (suggested to him by Ostrogradskiĭ) for residues and for the sum of the residues of a function in a given domain, usually a rectangle (Sections 5.2 and 5.3 above). He gives an elementary proof of the residue theorem for a rectangle, using the device that we have attributed to Ostrogradskiĭ (Section 5.4 above), and he obtains several formal properties of residues (Sections 5.6 and 5.7 above).

In one of his 1826 notes he relates the behaviour of a function $f(z)$ for large values of $|z|$ with that of $f(1/s)$ for small values of $|s|$ (Section 5.6 above); it is tempting to speculate that this was when he began to realise that there might be advantages in working with circles rather than with rectangles. In any case, his next step is to examine how residues are transformed under a change of variable (Section 5.7 above), and he uses his results here to obtain versions of his basic formulae in terms of polar coordinates. In particular, he writes down the residue theorem for a truncated sector of a circle (Section 5.8 above). Some of his formulae here look as if he is integrating round a circle; in fact, he has not yet quite reached this point of view, as is witnessed by his introduction of special conventions, according to which the centre of the circle is treated as if it were a boundary point, presumably because, under the transformation he uses, the centre is the image of a point at infinity.

In 1827 Cauchy begins to reap the benefit of his change-over from rectangles to circles. In 1826 he had tried to show that if a function $f(z)$ vanishes at infinity, and $zf(z)$ reduces to a constant F when z becomes infinite through real or purely imaginary values, then the sum of all the residues of $f(z)$ is equal to F (Section 5.5 above); the proof he gives there is inadequate in several respects (see also Section 7.8 below). He now remarks that this proposition requires a much more precise interpretation than he had given before; assuming that the only singularities of $f(z)$ are infinities (poles), what he does is to introduce a sequence of concentric circles tending to infinity and avoiding the poles, and to specify the behaviour of the function on these circles (Section 5.10 above). At this stage he abandons his special treatment for the centres of circles, and uses geometrical language much more freely than he had done before; he is now quite clearly thinking in terms of integrals round circles (integrals round general closed curves were to come a little later).

Among the applications he makes of these results is an important theorem on the expansion of a function whose only singularities are isolated poles as a series of rational functions (Section 5.11 above); what he achieves here may be regarded as a first step towards the theorems proved by Mittag-Leffler in 1884. In 1829 Cauchy takes a look at the problem of expressing a given function, having no infinities, as a product of linear factors, and succeeds in making a little progress in the direction of the Weierstrass product theorem for entire functions.

In 1831 he presented two important memoirs to the Turin academy; these were then issued in lithographed form. In the first of these, to which we shall return in Section 7.6, he gives by far his simplest proof of the result that was to become known as 'Cauchy's formula' for the value of an analytic function at a point inside a circle in terms of an integral round the circumference (Section 6.9 above).

In our present context the second Turin memoir is the more important one; in this he generalises 'Cauchy's theorem' and the residue theorem to the case of a domain bounded by a simple closed curve. He begins by introducing a correction term, very much as he had done in the footnotes to the 1814 memoir, using as before the double-integral idea and a generalised form of the Cauchy–Riemann equations; he then states, with no more than a hint at the proof, that the correction term is equal to $2\pi i$ times the sum of the residues of the function at its poles inside the contour (Section 6.14 above). He then goes on to express the correction term as an integral round the contour; his argument here is very close to that normally used to prove the divergence theorem in the plane (Section 6.15 above).

Although he does not quote the 1825 memoir explicitly, some of the ideas introduced in it reappear in the present context; these include the geometrical language, the integration of a function along a curve in the complex plane, and the device we have attributed to Ostrogradskiĭ.

At this point, with the proof of the general residue theorem, Cauchy's work has achieved a certain finality; the theory that he has been developing since 1814 has culminated in a simple and general proposition covering all his main results in this context. Later in the memoir he uses his general theorem to obtain information about the zeros of a function inside the contour; we defer the discussion of these applications to Section 7.6.

It is perhaps appropriate at this point to sum up one aspect of the development of the theory in the following statements; his first results were presented in terms of real integrals between real limits (1814), then came integrals of complex functions over real intervals (from c. 1817 onwards), integrals of complex functions along (originally monotone) paths in the complex plane (1825), then integrals round circles (1827), and finally integrals round general simple closed curves (1831).

7.6. Throughout the period with which we are concerned, Cauchy maintained his interest in the solution of equations, both algebraic and transcendental. Among the types of problems that he investigated were the existence of solutions, the localisation of the roots of equations, explicit formulae and series expansions for roots, and methods of approximation.

In some of these directions his work led to important results in complex function theory.

His first results on localisation appeared in his paper [1813] on the determination of the number of real roots of an algebraic equation in a given interval. In 1817 he published two proofs of the fundamental theorem of algebra on the existence of roots of algebraic equations, and in 1821 he extended his proof to equations with complex coefficients (Section 3.11 above); Argand had already outlined a proof for this case in 1815.

Although he had made substantial contributions to the theory of permutations, Cauchy appears never to have worked on the problem of solving algebraic equations in terms of radicals. In his memoir [1819], which was never published in full, although a brief summary of it appeared in 1824, he explains that he has accepted the validity of Ruffini's proof that the general equation of the fifth degree is insoluble by radicals; he goes on to announce that he has found a way of expressing each root of an algebraic equation as a definite integral whose integrand is a rational function. He published some formulae of this kind in 1823, not only for the roots themselves, but also for expressions of the form

$$\varphi(x_1) + \varphi(x_2) + \cdots + \varphi(x_m),$$

where $\varphi(x)$ is a given function, and $x_1, x_2, \ldots, x_m$ are the roots of an equation $F(x) = 0$ lying in a specified domain (Section 3.11 above).

He returned to formulae of this kind in 1827, now basing them on the theory of residues, and showing that the formulae continue to hold, when properly interpreted, in cases where multiple roots are present (Section 5.9 above).

In 1770 Lagrange gave a series solution for equations of the form

$$\alpha - x + \varphi(x) = 0,$$

where $\varphi(x)$ is a polynomial (Section 6.2 above), and later extended it to the transcendental equation

$$t = u - c \sin u,$$

usually known as Kepler's equation, which arises in the theory of planetary orbits; the series he obtained was to become known as the Lagrange series (Section 6.3 above). Laplace proposed a generalisation of it in 1779, but in 1788 Paoli pointed out that Laplace's statement required modification in certain cases (Section 6.4 above).

All this work was purely formal; the first serious study of the convergence of the series was made by Laplace in 1825, possibly as the result of a remark by Cauchy (Section 6.5 above). Cauchy then began an investigation of the problem. In an 1827 paper he used his results on the expression of the higher derivatives of a function as contour integrals to determine the asymptotic behaviour of the terms of the Lagrange series (Section 6.6 above), and in another paper of the same year he found an explicit expression for the remainder of the series after a finite number of terms (Section 6.7 above).

It must have been about this time that Cauchy realised that his contour integral expressions for functions and their derivatives constituted an extremely powerful set of tools, and he begins to exploit them fully in his first Turin memoir of 1831. His first main result is his proof that the Maclaurin series of an analytic function is convergent inside the circle, with centre at the origin, whose radius is the distance from the origin to the nearest singularity of the function (Section 6.9 above). He then obtains estimates for the terms of the series and for its remainder after n terms, thus founding what he calls the 'calculus of limits', later to become known as the method of majorants.

He goes on to apply a similar technique to obtain an explicit expression in the form of a contour integral for sums such as

$$F(y_1) + F(y_2) + \cdots + F(y_m),$$

where F is a given function and $y_1, y_2, \ldots, y_m$ are those roots of an equation

$$f(x, y) = 0$$

that lie in the domain $|y - b| < Z$. In particular, this enables him to give formulae for the number of roots in the domain, and for a single root if it happens to be the only one (Section 6.10 above). He uses his results to obtain expansions for these sums as power series in x, and also expansions for more general expressions of the form

$$F(x, y_1) + F(x, y_2) + \cdots + F(x, y_m).$$

He gives estimates for the coefficients of the series (Section 6.11 above). He shows that his results can be specialised to give the Lagrange series, and he calculates the radius of convergence for the series arising in Kepler's problem (Section 6.13 above). Finally, he proves a result that was to be rediscovered by Rouché in 1862.

In his second Turin memoir of 1831 he obtains contour-integral formulae for expressions of the form

$$F(z_1) + F(z_2) + \cdots + F(z_m),$$

where $z_1, z_2, \ldots, z_m$ are those roots of an equation $f(z) = 0$ that lie inside a given simple closed contour (Section 6.16 above). In particular, by taking $F(z) = 1$, he has a contour-integral formula for the number of roots lying inside the contour: he transforms this expression into another form, which enables him to express m in terms of what he calls the *index* of $f(z)$ with respect to the contour (Sections 6.17 and 6.18 above). For his purposes the index plays a role similar to that played by what was later to be called the winding number; he builds up an elaborate theory, which he calls the 'calculus of indices'.

As a first application of these ideas, he proves the general form of Rouché's theorem (as it is usually called, although Rouché only considered the case of a circle) for a simple closed contour; this states that if

$$f(z) = \Pi(z) + \varpi(z),$$

where $|\varpi(z)| < |\Pi(z)|$ on the contour, then $f(z)$ and $\Pi(z)$ have the same number of zeros inside the contour (Section 6.19 above). Writing z_1, $z_2, \ldots, z_m$ for the zeros of $f(z)$ and $\zeta_1, \zeta_2, \ldots, \zeta_m$ for those of $\Pi(z)$, Cauchy obtains an infinite series for expressions of the form

$$F(z_1) + F(z_2) + \cdots + F(z_m) - F(\zeta_1) - F(\zeta_2) - \cdots - F(\zeta_m),$$

which involves the residues of the powers of $\varpi(z)/\Pi(z)$, and he shows that the Lagrange series and other related series appear as special cases.

Cauchy's second Turin memoir has never appeared in print in its original form, although an Italian translation was published in Modena in 1839. That Cauchy never thought it worth republishing may be connected with the fact that in 1833 he recast his calculus of indices in a form that no longer involved the use of contour integrals (Section 6.20 above); it seems that it was this aspect of the paper that he regarded as important, rather than the results dealing with complex integration round general simple closed contours (Section 6.21 above).

7.7. We now take a look at the role played by principal-value integrals in Cauchy's work. He introduced the notion in an informal way in Part II of his 1814 memoir (Section 2.9 above); many of the definite integrals evaluated there have to be interpreted in this sense. He gave a formal definition in 1822 (Section 3.8 above); at the same time he defined what he called the 'general value' of the integral of a function with an infinity in the range of integration, but it seems that he never found any serious use for it.

In the 1814 memoir he extends his discussion of the correction term (essentially the real or imaginary part of the residue) to cover cases where the function has an infinity on the boundary of the domain, and shows that, in the case where the pole is simple, the corresponding expression has to be halved (Section 2.12 above). He repeats this remark in his revised expositions of the theory in 1822–23 (Section 3.6 above). In 1822, when he is expressing a definite integral in terms of the residues of the integrand, he explicitly links an infinity on the boundary with a principal-value integral over that part of the boundary (Section 3.7 above).

In the 1825 memoir, when he is investigating the difference between integrals along neighbouring paths, he allows a simple pole to lie on one of the paths, so that the integral along that path has to be interpreted as a principal value; this leads to a factor $\pm\pi i$ in the correction term. He deduces that if the two paths pass on opposite sides of the pole, the factor will be $\pm 2\pi i$ (Section 4.5 above).

When he comes to consider multiple poles, he begins with paths that pass on either side of the pole (Section 4.6 above); he subsequently examines what happens if a multiple pole lies on one of the paths, and finds that, except in special cases, the correction term becomes infinite (Section 4.7 above). There is a hint here that he is somewhat disappointed to find that principal values do not in general give useful results in this case.

After giving a formal definition in 1826 of the notion of residue, he remarks that the sum of the residues of a function over a domain can be made into what we might call an additive function of rectangles, provided that a pole lying on a side of the rectangle contributes one-half of its residue to the sum, and a pole at one of the vertices contributes one-quarter (Section 5.3 above). In a final effort, he interprets the contribution to the integral over a pole at one of the vertices as what we have called a 'biprincipal value', and he examines the conditions for a multiple pole to give a finite result in this case (Section 5.4 above).

From this point on he seems to lose interest in the contributions from poles on the boundary; there is one last mention of halving the residue in certain cases, but it seems to have been a routine remark (Section 5.8 above), and it does not seem that it is intended to be taken very seriously.

7.8. In this section we trace Cauchy's successive attempts to find what conditions should be imposed upon the behaviour of functions at infinity to ensure the validity of his arguments. Initially, his conditions were not strong enough to warrant the conclusions he draws from them. Eventually cases emerged where they led to erroneous results; he made several attempts to improve on them before he reached a satisfactory conclusion, about 1827.

In the 1814 memoir he tacitly assumes that if $P(X, z)$ vanishes when $X = +\infty$, then the same is true for

$$\int_0^Z P(X, z)\,\mathrm{d}z$$

(Section 2.5 above); for finite Z this would often be legitimate, but he sometimes makes the same assumption when $Z = +\infty$ (Sections 2.7 and 2.13 above). The same point arises in 1823 in the course of his revised proof of his basic theorem in the *Calcul infinitésimal* (Section 3.6 above).

One of his standard results is that if $f(x)$ has infinities at $x_1, x_2, \ldots, x_n$ in the upper half-plane, with residues $f_1, f_2, \ldots, f_m$, then

$$\int_{-\infty}^{\infty} f(x)\,\mathrm{d}x = 2\pi\mathrm{i}(f_1 + f_2 + \cdots + f_m),$$

under the hypothesis that $f(x + y\mathrm{i})$ vanishes for all y when $x = \pm\infty$ and for all x when $y = +\infty$ (Section 3.9 above). This was to be his usual hypothesis for some time. However, in another 1823 publication, a case occurs where his usual conditions would have led to an erroneous result, although he reaches the correct answer by an *ad hoc* argument, without drawing the reader's attention to the inconsistency (Section 3.9 above).

His standard conditions reappear in the 1825 memoir (Section 4.9 above), and are even somewhat weakened to cover cases where the function has an infinity of singularities (Section 4.10 above). However, he does recognise the inconsistency that had emerged in 1823, and suggests another *ad hoc* device to get round it (Section 4.11 above). He is obviously puzzled, remarking that his usual assumption, that if $f(X + y\mathrm{i})$ vanishes when $X = +\infty$ then the same will be true of

$$\int_{y_0}^{Y} f(X + y\mathrm{i})\,\mathrm{d}y,$$

does not always hold.

In a later paper of 1825 he adds to his usual hypothesis the condition that $xf(x)$ has a finite limit F when $x \to \pm\infty$ (apparently through real values), and concludes that, if $f(x)$ has no infinities in the upper half-plane, then

$$\int_{-\infty}^{\infty} f(x)\,\mathrm{d}x = -\pi F\mathrm{i}$$

(Section 4.14 above). His argument still leaves something to be desired from the point of view of rigour, and it is not hard to find examples where it would not work.

In the long list of special results in the second part of this paper, he adds a footnote indicating which of them have to be modified to take account of this change.

In one of his early notes in the *Exercices de mathématiques* in 1826, he strengthens his hypothesis a little more, adding the condition that $(x + yi)f(x + yi)$ has the limit F when y becomes infinite (Section 5.5 above). Even so, his conditions are still not quite adequate.

In 1827, in the important paper in which he introduced a sequence of circles tending to infinity, he finally stated his condition in the form that $zf(z)$ should reduce to the constant F when $|z|$ goes to infinity with z on this sequence of circles, and so obtained a rigorous proof of his results (Section 5.10 above). This was one of the rewards for the simplification of theory achieved through the change from rectangles to circles. The precise statement of the conditions made here ended the difficulty, which disappears altogether at this stage.

7.9. In this final section we shall try to evaluate the contributions that Cauchy made from 1814 to 1831 to the development of complex function theory and to some auxiliary topics of analysis, and to put them as nearly as possible in chronological order.

It has become clear in the course of this work that Cauchy usually thought of a function as being given by a formula or an equation in its range of definition; where he differed from his predecessors was that he was prepared to consider functions that were defined only for a restricted range of values of the independent variable, or were defined by different equations for different ranges; it was in this sense that he rejected in this context the principle of the generality of analysis (Section 7.2). As a result, he tacitly assumed that a continuous function could be differentiated within its range of definition, even when he was dealing with functions of a complex variable. Even when he introduced the additional hypothesis of continuous differentiability, as he did in the republished version of his first Turin memoir in 1841, he did not do so for the reasons that would be adduced today.

He insisted from the beginning that the definite integral should be regarded as being in some sense a sum of differentials, until in 1823 he gave his formal definition of the integral of a continuous function over a finite interval as a limit of approximating sums (Section 7.3). This enabled him not only to get rid of some long-standing paradoxes, but also to establish the existence of a primitive function for an arbitrary continuous integrand.

In 1822 he introduced the notion of a principal-value integral; it proved to be useful in some special circumstances, but the idea faded into the background in his later papers, not being in the end as useful as he may perhaps have expected (Section 7.7).

His 1814 memoir was planned as an examination of the validity of the use of imaginary substitutions for the evaluation of definite integrals. In the first part of the memoir he obtained results that can be seen to be equivalent to 'Cauchy's theorem' for a class of curvilinear quadrilaterals; he did so by combining two techniques that were already known, the inversion of the order of integration in a repeated integral and the use of the 'Cauchy–Riemann equations'. In the second part he examined what happened when the function had singularities in the domain, so that the repeated integrals used in the first part might fail to be equal; he showed that the difference between them could be reduced to a 'singular integral', which could be thought of as a non-zero integral over an infinitely small interval. The results he obtained can be seen to be equivalent to the residue theorem for functions having only simple poles, for the same class of domains as in the first part of the memoir. His proofs involved some complicated analytic manipulations, and it is clear that at this stage he did not fully understand what was going on; he had probably attained a better grasp of the situation by 1822, and certainly by 1825. Nevertheless, his results were adequate for the applications he was making at the time.

Between 1814 and 1825 he made some technical improvements in his treatment of the subject; the most important of these was the use of complex integrands instead of real-valued ones. His proof for the special case of a rectangle was substantially simplified by 1823, although it still depended on the use of singular integrals (Section 7.5).

In 1825, there appeared Cauchy's famous memoir on integrals between imaginary limits, which has been regarded as his most important contribution during this period to complex analysis. He himself hardly ever quoted its results, although some of its ideas can be recognised in his second Turin memoir. It also contains some technical improvements, including a full treatment of multiple poles, and the device of splitting a function in the neighbourhood of a pole into a simple rational expression and a regular function (a device that we have attributed to Ostrogradskiĭ); the latter enabled him to dispense with the use of singular integrals. It was also in this memoir that he began to introduce some geometrical language in the complex plane (Section 7.5). These improvements were carried on into his later work, including the long series of notes in the *Exercices de mathématiques* from 1826 onwards.

In 1826 he gave a formal definition for the residue of a function relative to a pole, and established some of the properties of residues. At this time he still treated the residue theorem for a rectangle as fundamental, deriving it for other domains by a coordinate transformation (Section 7.5). In 1827 he began the shift from rectangles to circles; this move enabled him to clear up the difficulties about the conditions to be imposed on functions at infinity that had been giving him trouble since 1823 (Section 7.8).

It was at this time that he began to express some of his results in terms of integrals round closed curves, although for the moment these were only circles. The gain in simplicity and clarity became manifest immediately, and it is particularly noticeable in his first Turin memoir of 1831, where he worked in terms of circles from the beginning (Section 7.5). In the second Turin memoir of the same year, he extended his results to more general simple closed contours, using an argument resembling a now familiar proof of the divergence theorem in the plane (Section 7.5).

Up to about 1825 the main applications that he made of his results were to the evaluation of definite integrals; from then onwards these were supplemented by applications to the summation of infinite series.

Another application that played an increasing role in Cauchy's work from 1819 onwards was the expression of the roots of algebraic and transcendental equations as contour integrals; in particular, he regarded this as a way of getting round the impossibility of solving higher-degree algebraic equations by radicals (Section 7.6). He also became interested in expressions for the roots of equations as infinite series, such as that introduced by Lagrange in 1770–71. In 1827 he began an investigation of the convergence of the Lagrange series; this culminated in the two Turin memoirs (1831), and led him on to examine more general series arising in the theory of implicit functions (Section 7.6). It was almost as a by-product of this work that, in the first Turin memoir, he proved the convergence of the Maclaurin series of a function up to the singularity nearest to the origin (Section 7.5); it was in this context that he created what he called the 'calculus of limits', later known as the method of majorants.

In the second Turin memoir (1831) he showed that the number of zeros of a function $f(z)$ inside a given closed curve can be expresed as a contour integral; he transformed this expression into a form depending on what he called the index of $f(z)$ with respect to the contour, closely related to what was later to be called the winding number. He developed these results into what he called a 'calculus of indices' (Section 7.6).

It is a recognised phenomenon in the history of mathematics that, when the first proof of a major result is extremely complicated, it is usually the task of the pioneer's successors to improve and simplify the proofs. Cauchy's early work on complex function theory is an exception to this rule; it was he himself who achieved the remarkable simplifications that converted his original complicated proofs into the elegant and satisfying techniques of his later papers. One reason for this was that he was working practically alone; except for a few contributions made by Ostrogradskiĭ about 1825, nobody was showing serious interest either in his methods or in his general results.

A further consequence of this lack of outside interest was Cauchy's failure to grasp how fertile some of his results were to become. For instance, he undervalued the importance of his 1825 memoir on integrals along paths in the complex plane; he never quoted its general results, only using some technical devices from it that he had introduced along the way. Again, he never published in printed form the full results of his second Turin memoir, in which he had extended the residue theorem to general simple closed contours; he evidently regarded his calculus of indices as the principal contribution of that memoir, and he had found subsequently that it could be developed without reference to contour integration.

Nevertheless, the whole body of work done by Cauchy on complex function theory between 1814 and 1831 is a remarkable achievement for one man, who was essentially working in isolation.

References

d'Alembert, J. le R. (1752). *Essai d'une nouvelle théorie de la résistance des fluides.* David, Paris.
—— (1761a). Recherches sur les vibrations des cordes sonores. *Opuscules mathématiques,* **1**, 1–64.
—— (1761b). Remarques sur les loix du mouvement des fluides. *Opuscules mathématiques,* **1**, 137–68.
—— (1768a). Sur un paradoxe géométrique. *Opuscules mathématiques,* **4**, 62–5.
—— (1768b). Sur l'équilibre des fluides. *Opuscules mathématiques,* **5**, 1–40.
Argand, J. R. (1806). *Essai sur une manière de représenter les quantités imaginaires dans les constructions géométriques,* 2nd edn, Paris, 1874.
—— (1813). Essai sur une manière de représenter les quantités imaginaires, dans les constructions géométriques. *Annales de Math.* (Gergonne), **4**, 133–47 (1813).
—— (1815). Réflexions sur la nouvelle théorie des imaginaires, suivies d'une application à la démonstration d'un théorème d'analise. *Annales de Math.* (Gergonne), **5**, 197–209 (1815).
Badolati, E. (1977). Sintesi storica della risoluzione dell' equazione di Keplero per mezzo della serie di Lagrange. *Rend. Accad. Sci. Fis. Mat. Napoli,* **(4)43**, 128–39.
Belhoste, B. (1991). *Augustin-Louis Cauchy: A Biography.* Springer-Verlag, New York.
Bernoulli, Johann (1702). Solution d'un problème concernant le calcul intégral, avec quelques abrégés par rapport à ce calcul. (Extract of letter written from Groningen, 5 August 1702.) *Mém. Acad. Sci. Paris,* 1702, 289ff; *Opera Omnia,* **1**, 393–400 (1742).
Berresford, G. C. (1981). Cauchy's theorem. *Amer. Math. Monthly,* **88**, 741–4.
Bidone, G. (1812). Mémoire sur diverses intégrales définies. *Mém. Acad. Imper. Turin* (1811–12), 231–344 (1813). (Read 28 May 1812.)
Bottazzini, U. (1986). *The Higher Calculus: a History of Real and Complex Analysis from Euler to Weierstrass.* Springer-Verlag, New York.
—— (1990). Editor's introduction in: A. L. Cauchy, *Cours d'analyse de l'Ecole Royale Polytechnique.* Cooperativa Libraria Universitaria Editrici, Bologna.
Brill, A. and Noether, M. (1894). Die Entwicklung der Theorie der algebraischen Functionen in älterer und neuerer Zeit. *Jahresber. Deutsch. Math. Ver.* **3**, 107–565.
Casorati, F. (1868). *Teorica delle funzioni di variabili complesse.* Fusi, Pavia.

Cauchy, A. L. (1813). Mémoire sur la détermination du nombre des racines réelles dans les équations algébriques. *J. Ecole Polytechnique Paris*, **10** (cahier 17), 457–548 (1815); *Oeuvres complètes* **(2)1**, 170–257 (1905). (Commun. 17 May, 18 October and 22 November 1813.)

—— (1814). Mémoire sur les intégrales définies. *Mémoires présentés par divers savans à l'Académie Royale des Sciences*, **(1)1**, 601–799 (1827); *Oeuvres complètes* **(1)1**, 329–506 (1882). (Read 22 August 1814; MS submitted for printing 14 September 1825.)

—— (1815). Théorie de la propagation des ondes à la surface d'un fluide pesant d'une profendeur indéfinie. *Mémoires présentés par divers savans à l'Académie Royale des Sciences* **(1)1**, 3–312 (1827). *Oeuvres* **(1)1**, 5–318 (1882). (Submitted for Academy prize, 26 December 1815.)

—— (1817a). Sur les racines imaginaires des équations. *Bull. Soc. Philomat. Paris*, 1817, 5–9, *Oeuvres* **(2)2**, 210–16 (1958) (Commun. 23 December 1816.)

—— (1817b). Seconde note sur les racines imaginaires des équations. *Bull. Soc. Philomat. Paris*, 1817, 161–4; *Oeuvres* **(2)2**, 217–22 (1958). (Commun. 13 October 1817.)

—— (1818). Note sur l'intégration d'une classe particulière d'équations différentielles. *Bull. Soc. Philomat. Paris*, 1818, 17–20; *Oeuvres* **(2)2**, 233–7 (1958). (Commun. 19 January 1818.)

—— (1819). Sur la résolution analytique des équations de tous les degrés, par le moyen des intégrales définies. *Mém. Acad. Sci. Paris*, 4 (1819–20), xxvi–xxix (1824); *Oeuvres* **(1)2**, 9–11 (1908). (Commun. 22 November 1819.)

—— (1820). Sur les racines imaginaires des equations. *J. Ecole Polytechnique Paris*, **11** (cahier 18), 411–16; *Oeuvres* **(2)1**, 258–63 (1905). (Commun. 13 October 1817.)

—— (1821a). *Cours d'analyse de l'Ecole Royale Polytechnique, Ire Partie, Analyse Algébrique*, Paris; *Oeuvres* **(2)3** (1897).

—— (1821b). Mémoire sur l'intégration des équations linéaires aux différences partielles, à coefficients constants et avec un dernier terme variable. *Bull. Soc. Philomat. Paris*, 1821, 101–12; *Oeuvres* **(2)2**, 253–66 (1958). (Commun. 8 October 1821.)

—— (1822a). Sur le développement des fonctions en séries et sur l'intégration des équations différentielles ou aux différences partielles. *Bull. Soc. Philomat. Paris*, 1822, 49–54; *Oeuvres* **(2)2**, 276–82 (1958). (Commun. 22 January 1822.)

—— (1822b). Mémoire sur les intégrales définiés où l'on fixe le nombre et la nature des constantes arbitraires et des fonctions arbitraires que peuvent comporter les valeurs de ces mêmes intégrales quand elles deviennent indéterminées. *Bull. Soc. Philomat. Paris*, (Oct. 1822), 161–74; *Oeuvres* **(2)2**, 283–99 (1958). (Commun. 28 Oct. 1822.)

—— (1823a). Mémoire sur l'intégration des équations linéaires aux différentielles partielles et à coefficients constants. *J. Ecole Polytechnique Paris*, **12** (cahier 19), 510–92; *Oeuvres* **(2)1**, 275–357 (1905). (Commun. 16 Sept. 1822.)

—— (1823b). *Résumé des leçons données à l'Ecole Royale Polytechnique, sur le calcul infinitésimal*, Paris; *Oeuvres* **(2)4**, 5–261 (1899).

—— (1824). *Extrait d'un mémoire présenté à l'Académie Royale des Sciences, le 9 août 1824*, Paris; *Oeuvres* **(2)15**, 18–22 (1974). Reprinted as: Extrait d'un mémoire sur quelques séries analogues à la série de Lagrange, sur les fonctions symétriques et sur la formation directe des équations que produit l'élimination des inconnues entre des équations algébriques données. *Mém. Acad. Sci. Paris*, **9** (1826), 104–10 (1830); *Oeuvres* **(1)2**, 73–8 (1908).

—— (1825a). Sur les intégrales définies prises entre des limites imaginaires. *Bull. Sci. Math. Astr. Phys. Chim. (Bulletin de Férussac)* **3**, 214–21; *Oeuvres* **(2)2**, 57–65 (1958). (Commun. 28 Feb. 1825.)

—— (1825b). *Mémoire sur les intégrales définies prises entre des limites imaginaires*, Paris; *Oeuvres* **(2)15**, 41–89 (1974). (Commun. 28 Feb. 1825.) German translation under title 'Abhandlung über bestimmte Integrale zwischen imaginären Grenzen', with notes and commentary by P. Stäckel, in *Ostwalds Klassiker der exakten Wissenschaften*, **112** (1900).

—— (1825c). Mémoire sur les intégrales définies, où l'on donne une formule générale de laquelle se déduisent les valeurs de la plupart des intégrales définies déjà connues et celles d'un grand nombres d'autres (Première partie). *Annales de Math.* (Gergonne). **16**, 97–108; *Oeuvres* **(2)2**, 343–52 (1958).

—— (1825d). Rapport sur un mémoire de Barnabé Brisson sur l'"Intégration des équations linéaires aux différences finies ou infiniment petites." *Procès-Verbal Acad. Sci. Paris*, **8**, 223–6; *Oeuvres* **(2)15**, 560–5 (1974).

—— (1826a). Sur un nouveau genre de calcul analogue au calcul infinitésimal. *Exerc. Math.* **1**, 11–24; *Oeuvres* **(2)6**, 23–37 (1887). (Commun. 27 Feb. 1826.)

—— (1826b). De l'influence que peut avoir, sur la valeur d'un intégral double, l'ordre dans lequel on effectue les intégrations. *Exerc. Math.* **1**, 85–94; *Oeuvres* **(2)6**, 113–23 (1887). (Commun. 29 May 1826.)

—— (1826c). Sur diverses relations qui existent entre les résidus des fonctions et les intégrales définies. *Exerc. Math.* **1**, 95–113; *Oeuvres* **(2)6**, 124–45 (1887).

—— (1826d). Sur quelques formules relatives à la détermination du résidu intégral d'une fonction donnée. *Exerc. Math.* **1**, 133–9; *Oeuvres* **(2)6**, 169–76 (1887).

—— (1826e). Sur quelques transformations applicables aux résidus des fonctions, et sur le changement de variable indépendante dans le calcul des résidus. *Exerc. Math.* **1**, 167–76; *Oeuvres* **(2)6**, 210–20 (1887).

—— (1826f). Recherche d'une formule générale qui fournit la valeur de la plupart des intégrales définies connues et celle d'un grand nombre d'autres (Deuxième partie). *Annales de Math.* (Gergonne), **17**, 84–127; *Oeuvres* **(2)2**, 353–87 (1958).

—— (1826g). Sur les limites placées à droite et à gauche du signe $\mathscr{E}$ dans le calcul des résidus. *Exerc. Math.* **1**, 205–32; *Oeuvres* **(2)6**, 256–85 (1887).

—— (1827a). Usage du calcul des résidus pour déterminer la somme des fonctions semblables des racines d'une équation algébrique ou transcendante. *Exerc. Math.* **1**, 339–57; *Oeuvres* **(2)6**, 401–21 (1887).

—— (1827b). Mémoire sur divers points d'analyse. *Mém. Acad. Sci. Paris*, **8** (1825), 97–100, 101–129 (1829); *Oeuvres* **(1)2**, 29–32, 33–58 (1908). (Commun. 3 Sept. 1827 under title 'Règles de convergence de la série de Lagrange et d'autres séries du même genre'.)

—— (1827c). Mémoire sur le développement de $f(\zeta)$ suivant les puissances ascendantes de h, ζ étant une racine de l'équation $z - x - h\varpi(z) = 0$. *Mém. Acad. Sci. Paris*, **8** (1825), 130–8 (1829); *Oeuvres* **(1)2**, 59–66 (1908). (Commun. 3 Sept. 1827 under title 'Sur la détermination du reste de la série de Lagrange par une intégrale definie'.)

—— (1827d). Sur quelques propositions fondamentales du calcul des résidus. *Exerc. Math.* **2**, 245–76; *Oeuvres* **(2)7**, 291–323 (1889). (Commun. 5 Nov. 1827.)

—— (1827e). Sur le développement des fonctions d'une seule variable en fractions. rationnelles. *Exerc. Math.* **2**, 277–97; *Oeuvres* **(2)7**, 324–44 (1889). (Commun. 10 Dec. 1827.)

—— (1827f). Usage du calcul des résidus pour la sommation ou la transformation des séries dont le terme général est une fonction paire du nombre qui représente le rang de ce terme. *Exerc. Math.* **2**, 298–314; *Oeuvres* **(2)7**, 345–62 (1889). (Commun. 17 Dec. 1827.)

—— (1827g). Sur un mémoire d'Euler, qui a pour titre *Nova methodus fractiones quascumque rationales in fractiones simplices resolvendi. Exerc. Math.* **2**, 315–16; *Oeuvres* **(2)7**, 363–5 (1889).

—— (1828). Sur les résidus des fonctions exprimées par des intégrales définies. *Exerc. Math.* **2**, 341–76; *Oeuvres* **(2)7**, 393–430 (1889). (Commun. 22 Jan. 1828.)

—— (1829a). Mémoire sur l'application du calcul des résidus à l'évaluation et à la transformation des produits composés d'un nombre infini de facteurs. *Bull. Sci. Math. Phys. Chim.* (*Bulletin de Férussac*), **12**, 202–5; *Oeuvres* **(2)2**, 84–7 (1958). (Commun. 17 and 31 Aug. 1829 under title 'Usage du calcul des résidus pour l'évaluation et la transformation des produits composés d'un nombre infini de facteurs'.)

—— (1829b). Sur la détermination du résidu intégral de quelques fonctions. *Exerc. Math.* **4**, 161–73; *Oeuvres* **(2)9**, 196–209 (1891). (Commun. 14 Sept. 1829.)

—— (1829c). Usage du calcul des résidus pour l'évaluation ou la transformation des produits composés d'un nombre fini ou infini de facteurs. *Exerc. Math.* **4**, 174–213; *Oeuvres* **(2)9**, 210–53 (1891). (Commun. 17 and 31 Aug. 1829.)

—— (1831a). Sur la mécanique céleste et sur un nouveau calcul qui s'applique à un grand nombre de questions diverses. *Bull. Sci. Math. Phys. Chim.* (*Bulletin de Férussac*) **15**, 260–69; *Oeuvres* **(2)2**, 158–68 (1958); also in *Exerc. d'Anal.* **2**, 41–9 (1841), under title 'Résumé d'un mémoire sur la mécanique céleste et sur un nouveau calcul appelé *calcul des limites*'; *Oeuvres* **(2)12**, 48–58 (1916).

—— (1831b). Sur les rapports qui existent entre le calcul des résidus et le calcul des limites et sur les avantages que présentent ces deux nouveaux calculs dans la résolution des équations algébriques ou transcendantes. *Bull. Sci. Math. Phys. Chim.* (*Bulletin de Férussac*), **16**, 116–19; *Oeuvres* **(2)2**, 169–72 (1958).

—— (1831c). Formules extraites d'un mémoire présenté le 27 novembre 1831 à l'Académie des Sciences de Turin. *Bull. Sci. Math. Phys. Chim.* (*Bulletin de Férussac*), **16**, 119–28; *Oeuvres* **(2)2**, 173–83 (1958).

—— (1831d). *Extrait du mémoire présenté à l'Académie de Turin, le 11 octobre 1831.* Lithograph, Turin (1833); also in *Exerc. d'Anal.* **2**, 50–108 (1841) under title 'Formules pour le développement des fonctions en séries'; *Oeuvres* **(2)12**, 58–112 (1916). Italian translation of the memoir by Paolo Frisiani and Gabrio Piola, under title 'Sulla meccanica celeste e sopra un nuovo calculo chiamato calcolo dei limiti'. *Opascoli matematici e fisici*, **2** (1834), with editorial notes. (There is an offprint of this translation in the library of St. John's College, Cambridge; in this the whole memoir occupies 127 pages, and there are 82 pages of notes.)

—— (1831e). *Mémoire sur les rapports qui existent entre le calcul des résidus et le calcul des limites, et sur les avantages qu'offrent ces nouveaux calculs dans la résolution des équations algébriques ou transcendantes*, Lithograph, Turin (1832–33). (Presented to Turin Academy, 27 November 1831.) *Oeuvres* **(2)15**, 182–261 (1974). Italian translation by Antonio Lombardi under title 'Memoria sui rapporti che esistono fra il calculo dei residui ed il calcolo dei limiti e sui vantaggi che offrono questi due nuovi calcoli nella soluzione delle equazioni algebraiche o transcendenti. *Mem. Soc. Ital. Sci.* (*Modena*), **22** (Parte Mat.), 91–183 (1839). (Received 23 October 1837.)

—— (1833a). *Calcul des indices des fonctions*, Lithograph, Turin (1833); also in enlarged form in *J. Ecole Polytechnique Paris*, **15** (cahier 25), 176–229 (1837); *Oeuvres* **(2)1**, 416–66 (1905). Italian translation by Antonio Lombardi under title 'Calcolo degli indizi delle funzioni', *Mem. Soc. Ital. Sci.* (*Modena*) **22**, 228–46 (1839).

—— (1833b). *Résumés analytiques*, Turin (1833); *Oeuvres* **(2)10**, 5–184 (1895).

—— (1839). Note sur l'intégration des équations différentielles des mouvements planétaires. *Exerc. d'Anal.* **1**, 27–32; *Oeuvres* **(2)11**, 43–50 (1913).

Clairaut, A. C. (1740). Sur l'intégration ou la construction des équations différentielles du premier ordre. *Mém. Acad. Sci. Paris*, 1740, 293–323.

—— (1743). *Théorie de la figure de la Terre, tirée des principes de l'hydrostatique*, David, Paris.

Eberlein, W. F. (1979). It's all in the family; is Cauchy's integral theorem trivial? *Rochester Math. Reports*, No. 1.

Ettlinger, H. J. (1923). Cauchy's paper of 1814 on definite integrals. *Ann. Math.* (2) 23 (1922), 255–70 (1923).

Euler, L. (1740). De infinitis curvis eiusdem generis seu methodus inveniendi aequationes pro infinitis curvis eiusdem generis. *Comment. Acad. Sci. Petrop.* **7** (1734–5), 174–89, 180–3 (1740); *Opera Omnia* **(1)22**, 36–56 (1936).

—— (1747). Sur les logarithmes des nombres négatifs et imaginaires. *Opera Postuma*, **1**, 269–81 (1862); *Opera Omnia* **(1)19**, 417–38 (1932). (Commun. Acad. Berlin, 7 Sept. 1747.)

—— (1748). *Introductio in analysin infinitorum*. Lausanne; *Opera Omnia* **(1)8** (1922).

—— (1757). Continuation des recherches sur la théorie du mouvement des fluides. *Mém. Acad. Berlin*, **11** (1755), 316–61 (1757); *Opera Omnia* **(2)12**, 92–132 (1954).

—— (1768a). *Institutiones calculi integralis*, Vol. 1. St. Petersburg. *Opera Omnia* **(1)11** (1913).

—— (1768b). Considerationes de traiectoriis orthogonalibus. *Novi Comment. Acad. Sci. Petrop.* **14** (1769), 46–71 (1770); *Opera Omnia* **(1)28**, 99–119 (1955). (Commun. 18 August 1768 o.s.) German translation by O. Neumann in *Ostwalds Klassiker der exaken Wissenschaften*, **261**, 101–127 (1983), under title 'Betrachtungen über orthogonale Trajektorien', with notes and commentary by A. P. Yushkevich.

—— (1775a). Nova methodus fractiones quascumque rationales in fractiones simplices resolvendi. *Acta Acad. Sci. Petrop.* 1780 (part I), 32–46 (1783); *Opera Omnia* **(1)6**, 370–83 (1921). (Commun. 14 August 1775 o.s.)

—— (1775b). De representatione superficiei sphaericae super plano. *Acta Acad. Sci. Petrop.*, 1777, 107–132 (1778); *Opera Omnia* **(1)28**, 248–87 (1955). (Commun. 4 Sept. 1775 o.s.) German translation by A. Wangerin in *Ostwalds Klassiker der exakten Wissenschaften*, **261**, 128–163 (1983), under title 'Über die Abbildung einer Kugelfläche in einer Ebene', with notes and commentary by A. P. Yushkevich.

—— (1777a). De integrationibus maxime memorabilibus ex calculo imaginariorum oriundis. *Nova Acta Acad. Sci. Petrop.* **7** (1789), 99–133 (1793); *Opera Omnia* **(1)19**, 1–44 (1932). (Commun. 20 March 1777 o.s.) German translation by O. Neumann in *Ostwalds Klaisiker der exakten Wissenschaften*, **261**, 164–209 (1983) under title 'Über höchst bemerkenswerte aus dem Imaginären – Kalkül herstammende Integrationen', with notes and commentary by A. P. Yushkevich.

—— (1777b). Supplementum ad dissertationem praecedentem circa integrationen formulae $\int z^{m-1} \partial z/(1 - z^n)$ casu, quo ponitur $z = v(\cos \varphi + \sqrt{-1} \sin \varphi)$. *Nova Acta Acad. Sci. Petrop.* **7** (1789), 134–48 (1793); *Opera Omnia* **(1)19**, 45–62 (1932).

—— (1777c). Specimen integrationis abstrusissimae hac formula $\int \partial x/(1 + x)\sqrt[4]{(2x - 1)}$ contentae. *Nova Acta Acad. Sci. Petrop.* **9** (1791), 98–117 (1795); *Opera Omnia* **(1)19**, 228–50 (1932). (Commun. 26 March 1777 o.s.)

—— (1777d). Integratio formulae differentialis maxime irrationalis, quam tamen per logarithmos et arcus circulares expedire licet. *Nova Acta Acad. Sci. Petrop.* **9** (1791), 118–26 (1795); *Opera Omnia* **(1)19**, 251–61 (1932). (Commun. 26 March 1777 o.s.)

—— (1777e). Ulterior disquisitio de formulis integralibus imaginariis. *Nova Acta Acad. Sci. Petrop.* **10** (1792), 3–19 (1797); *Opera Omnia* **(1)19**, 265–86 (1932). (Commun. 31 March 1777 o.s.; date corrected by A. P. Yushkervich.) German translation by E. Schuhmann in *Ostwalds Klassiker der exakten Wissenschaften*, **261**, 210–228 (1983), under title 'Eine weitere Untersuchung über imaginäre Integrale', with notes and commentary by A. P. Yushkevich.

—— (1777f). De integrationibus difficillimis, quarum integralia tamen aliunde exhiberi possunt. *Nova Acta Acad. Sci. Petrop.* **14** (1797–98), 62–74 (1805); *Opera Omnia* **(1)19**, 369–89 (1932). (Commun. 31 March 1777 o.s.; Date corrected as in [1777e].)

—— (1777g). Integratio succincta formulae integralis maxime memorabilis $\int \partial z/(3 \pm zz)\sqrt[3]{(1 \pm 3zz)}$. *Nova Acta Acad. Sci. Petrop.* **10** (1792), 20–26 (1797); *Opera Omnia* **(1)19**, 287–96 (1932). (Commun. 28 April 1777 o.s.)

—— (1777h). De formulis differentialibus angularibus maxime irrationalibus, quas tamen per logarithmos et arcus circulares integrare licet. *Institutiones calculi integralis*, **4**, 183–194 (1794); *Opera Omnia* **(1)19**, 129–40 (1932). (Commun. St. Petersburg Academy, 5 May 1777 o.s.)

—— (1777k). De insigni usu calculi imaginariorum in calculo integrali. *Nova Acta Acad. Sci. Petrop.* **12** (1794), 3–21 (1801); *Opera Omnia* **(1)19**, 345–68 (1932). (Commun. 3 Nov. 1777 o.s.)

—— (1781). De valoribus integralium a termino variabilis $x = 0$ usque ad $x = \infty$ extensorum. *Institutiones calculi integralis*, **4**, 337–45 (1794); *Opera Omnia* **(1)19**, 217–27 (1932). (Commun. St. Petersburg Academy, 30 April 1781 o.s.) German translation by E. Schuhmann in *Ostwalds Klassiker der exakten Wissenschaften*, **261**, 229–39 (1932), with notes and commentary by A. P. Yushkevich.

—— (1785). De resolutione fractionum transcendentium in infinitas fractiones simplices. *Opuscula analytica*, **2**, 102–137 (1785); *Opera omnia* **(1)15**, 621–60 (1927).

Falk, M. (1883). Extrait d'une lettre adressée à M. Hermite. *Bull. Sci. Math.* **(2)7**, 137–9 (1883).

Freudenthal, H. (1971). Cauchy, Augustin-Louis. *Dictionary of Scientific Biography*, **3**, 131–48.

Frullani, G. (1818). Sopra la dipendenza tra i differenziali delle fanzioni e gli integrali definiti. *Mem. Soc. Ital. Sci.* (*Modena*), **18**, 458–517 (1820). (Recd. 4 Feb. 1818.)

Gauss, C. F. (1799). Demonstratio nova theorematis omnem functionen algebraican rationalem integram unius variabilis in factores reales primi vel secundi gradus resolvi posse. Dissertation Helmstedt (1799); *Werke*, **3**, 1–30 (1866).

—— (1816). Theorematis de resolubilitate functionum algebraicarum integrarum in factores reales demonstratio tertia. *Comment, Soc. Roy. Sci.* (*Göttingen*), **3** (1814–15), 135–42 (1816); *Werke*, **3**, 57–64 (1866).

—— (1849). Beiträge zur Theorie der algebraischen Gleichungen. *Abh. Ges. Wiss. Göttingen*, **4** (1848), 3–34 (1850); *Werke*, **3**, 73–102 (1866).

Grattan-Guinness, I. (1990). *Convolutions in French mathematics*, 1800–1840. Three volumes. Birkhäuser, Basel.

Klein, F. (1926). *Vorlesungen über die Entwicklung der Mathematik im 19. Jahrhundert*, Vol. 1, Springer, Berlin.

Kline, M. (1972). *Mathematical Thought from Ancient to Modern Times*. Oxford University Press, New York.

Lacroix, S. F. (1798). *Traité du calcul différentiel et du calcul intégral*, Vol. 2. Duprat, Paris.

—— (1814). *Traité du calcul différentiel et du calcul intégral*, Vol. 2, 2nd edn. Courcier, Paris.

Lagrange, J. L. (1766). Solutions de différents problèmes du calcul intégral. *Miscellanea Taurinensia*, 3^2 (1762–1765), 179–380 (1766); *Oeuvres* **1**, 471–668 (1867).

—— (1770a). Nouvelle méthode pour résoudre les équations littérales par le moyen des séries. *Mém. Acad. Roy. Sci. Berlin*, **24** (1768), 251–326 (1770); *Oeuvres*, **3**, 5–78 (1869). (Read 18 Jan. and 5 April 1770.)

—— (1770b). Sur le problème de Kepler. *Mém. Acad. Roy. Sci. Berlin*, **25** (1769), 204–33 (1771); *Oeuvres*, **3**, 113–38 (1869). (Read 1 Nov. 1770.)

—— (1781). Sur la construction des cartes géographiques. *Nouv. Mém. Acad. Roy. Sci. Berlin*, 1779, 161–210 (1781); *Oeuvres*, **4**, 637–92 (1869).

—— (1804). Leçons sur le calcul des fonctions. *J. Ecole Polytechnique*, **5** (cahier 12), 1–324; *Oeuvres*, **10** (1884). (The version in the *Oeuvres* is reprinted from the second edition, published 1806.)

Laplace, P. S. (1779). Mémoire sur l'usage du calcul aux différences partielles dans la théorie des suites. *Mém. Acad. Roy. Sci. Paris*, 1777, 99–122 (1780); *Oeuvres*, **9**, 313–35 (1893). (Commun. 16 June 1779.)

—— (1780). Mémoire sur les probabilités. *Mém. Acad. Roy. Sci. Paris*, 1778, 227–332 (1781); *Oeuvres*, **9**, 381–485 (1893). (Commun. 19 July 1780.)

—— (1782). Mémoire sur les approximations des formules qui sont fonctions de très grands nombres. *Mém. Acad. Roy. Sci. Paris*, 1782, 1–88 (1785); *Oeuvres*, **10**, 209–91 (1894).

—— (1809). Mémoire sur divers points d'analyse. *J. École Polytechnique, Paris*, **8** (cahier 15), 229–65; *Oeuvres*, **14**, 178–214 (1912).

—— (1810). Mémoire sur les approximations des formules qui sont fonctions de très grands nombres et sur leur application aux probabilités. *Mém. Inst. France* (*première classe*), **10** (1809), 353–415 (1810); *Oeuvres*, **12**, 310–53 (1898). (Commun. 9 April 1810.)

—— (1811a). Sur les intégrales définies. *Nouv. Bull. Sci. Soc. Philomat. Paris*, **2**, 262–6 (1811); not in *Oeuvres*.

—— (1811b). Mémoire sur les intégrales définies et leur application aux probabilités, et spécialement à la recherche du milieu qu'il faut choisir entre les résultats des observations. *Mém. Inst. France* (*première class*), **11** (1810), 279–347 (1811); *Oeuvres*, **12**, 357–412 (1898). (Commun. 29 April 1811.)

—— (1812). *Théorie analytique des probabilités*. Courcier, Paris.

—— (1814a). *Essai philosophique sur les probabilités*. Courcier, Paris; *Oeuvres*, **7**, v–cliii (1886).

—— (1814b). *Théorie analytique des probabilités*, 2me édn. Courcier, Paris.

—— (1820). *Théorie analytique des probabilités*, 3me édn. Courcier, Paris; *Oeuvres*, **7** (1886).

—— (1825). Sur le développement en série du radical qui exprime la distance mutuelle de deux planètes, et sur le développement du rayon vecteur elliptique. *Connaissance des Temps*, 1828, 311–21 (1825); also in *Supplément au Ve, volume du Traité de mécanique celeste* (1827); *Oeuvres*, **5**, 469–89; also in *Mém. Acad. Sci. Paris*, **6** (1823), 61–80 (1827), under title 'Sur le développement des coordonnées elliptiques'; *Oeuvres*, **12**, 549–66 (1898).

Laurent, P. A. (1843). Extension du théorème de M. Cauchy relatif à la convergence du développement d'une fonction suivant les puissances ascendantes de la variable. Report by Cauchy in *C. R. Acad. Sci. Paris*, **17**, 938–42; in Cauchy's *Oeuvres* **(1)8**, 115–17 (1893).

Legendre, A. M. (1811). *Exercices du calcul intégral sur divers ordres de transcendantes et sur les quadratures*, Vol. 1. Courcier, Paris.

—— (1814a). *Exercices du calcul intégral sur divers ordres de transcendantes et sur les quadratures*, quatrième partie. Courcier, Paris.

—— (1814b). Extrait du Procès-Verbal de la Séance de la Classe des Sciences Physiques et Mathématiques du lundi 7 novembre 1814. *Mém. Div. Savans Acad. Sci. Paris*, **1**, 601–10 (1827); in Cauchy's *Oeuvres* **(1)1**, 321–7 (1882).

Libri. G. (1822). Mémoire sur divers points d'analyse. *Mem. R. Accad. Sci. Torino*, **28**, 251–80 (1824). (Commun. 14 July 1822.)

—— (1825). Mémoire sur quelques formules générales d'analyse. *J. Reine Angew. Math.* **7**, 57–67 (1831). (Also submitted to Acad. Sci. Paris in 1823 and 1825.)

Lindelöf, E. (1905). *Le calcul des résidus et ses applications à la théorie des fonctions*. Gauthier-Villars, Paris.

Osgood, W. F. (1901). Allgemeine Theorie der analytischen Funktionen (a) einer und (b) mehrerer complexen Grössen. *Encykl. Math. Wiss.*, **IIB1**, Band 2, Teil 2, 1–114.

Ostrogradskiĭ, M. V. (1828). Note sur les intégrales définies. *Mém. Acad. Imp. Sci. St. Pét.* **(6)1**, 117–22 (1831). (Commun. 29 October 1828 o.s.)

Paoli, P. (1788). Ricerche sulle serie. *Mem. Mat. Fis. Soc. Ital.* (*Modena*), **4**, 429–54 (1788).

Parseval, M. A. (1805). Mémoire sur les séries et sur l'intégration complète d'une équation aux différences partielles linéaires du second ordre à coefficiens constans. *Mém. Div. Sav. Inst. France*, **(1)1**, 638–48 (1805).

Poisson, S. D. (1811a). Sur les intégrales définies. *Nouv. Bull. Sci. Soc. Philomat. Paris*, **2**, 243–52/375–80.

—— (1811b). Abstract of Laplace [1811b] under title 'Mémoire sur les fonctions génératrices, les intégrales définies, et leur application aux probabilités'. *Nouv. Bull. Sci. Soc. Philomat., Paris*, **2**, 360–64.

—— (1813). Mémoire sur les intégrales définies. *J. Ecole Polytechnique, Paris*, **9** (cahier 16), 215–46.

—— (1814). Abstract of Cauchy [1814]. *Bull. Soc. Philomat., Paris*, 1814, 185–8; also in Cauchy's *Oeuvres* **(2)2**, 194–8 (1958).

—— (1820). Suite du mémoire sur les intégrales définies. *J. Ecole Polytechnique, Paris*, **11** (cahier 18), 295–341.

—— (1822). Mémoire sur les intégrales définies et sur la sommation des séries. *Bull. Soc. Philomat., Paris*, 1822, 134–9.

—— (1823a). Expression des fonctions par des séries périodiques. *J. Ecole Polytechnique, Paris*, **12** (cahier 19), 432–64.

—— (1823b). Nouvelles formules relatives aux intégrales définies. *J. Ecole Polytechnique, Paris*, **12** (cahier 19), 481–501.

Pringsheim, A. (1898). Irrationalzahlen und Convergenz unendlicher Processe. *Encykl. Math. Wiss.*, **1**, 47–146 (IA3).

Rouché, E. (1862). Mémoire sur la série de Lagrange. *J. Ecole Polytechnique, Paris*, **22** (cahier 39), 193–224.

Smithies, F. (1986). Cauchy's conception of rigour in analysis. *Arch. Hist. Exact. Sci.*, **36**, 41–62.

Stäckel, P. (1900). Integration durch das imaginäre Gebiet. *Bibliotheca Math.* **(3)1**, 109–28.

—— (1901). Beitrag zur Geschichte der Funktionentheorie im 18. Jahrhundert. *Bibliotheca Math.* **(3)2**, 111–21.

Terracini, A. (1957). Cauchy a Torino. *Rend. Sem. Mat. Torino*, **16**, 159–203.

Timchenko, I. (1899). Osnovaniya teoriya analiticheskikh funktsii, 1. Istorichkeskaya svedeniya. [Foundations of analytic function theory, 1. Historical account.] Odessa.

Truesdell, C. A. (1954). Rational fluid mechanics, 1687–1765. In L. Euler: *Opera omnia* **(2)12**, i–cxxv.

Valson, C. A. (1868). *La vie et les travaux du baron Cauchy*, 2 volumes. Paris. Reprinted in 1 volume, Gauthier-Villars; Paris (1970).

Vernier, H. V. (1824). Recherches sur la sommation de la série de Taylor et sur les intégrales définies. *Annales de Math.* (Gergonne), **15**, 165–89.

Yushkevich, A. P. (1965). O neopublikovannikh rannikh rabotakh M. V. Ostrogradskogo. [On unpublished early works of M. V. Ostrogradskiĭ.] *Istoriko-Matematicheskie Issledovaniya*, No. 16, 11–48.

Notation index

Author index

d'Alembert, J. le R. 2, 7–9, 23, 36, 42, 188, 190
Argand, J. R. 80, 84, 99, 111, 196

Badolati, E. 151
Belhoste, B. 4, 23, 98, 188
Bernoulli, Johann 11, 23
Berresford, G. C. 91
Bessel, F. W. 1
Bidone, G. 3, 51
Bottazzini, U. 4
Brill, A. 2, 4
Brisson, B. 85, 192

Casorati, F. 4
Cauchy, A. L. *passim*
Clairaut, A. C. 6–8, 12, 23

Darboux, G. 85, 102, 104

Eberlein, W. F. 91
Ettlinger, H. J. 29
Euler, L. 2, 7, 9–14, 16, 19–23, 25, 31, 50, 56, 138–9, 189–90

Falk, M. 90–1
Fourier, J. B. J. 75–6
Fresnel, A. J. 14
Freudenthal, H. 4, 146
Frisiani, P. 157–8, 160, 169, 188
Frullami, G. 73, 129

Gauss, C. F. 1, 79–80, 159
Grattan-Guinness, I. 4, 5, 139

Hadamard, J. 82

Klein, F. 2, 4
Kline, M. 4, 80

Lacroix, S. F. 24, 36, 54, 139, 188
Lagrange, J. L. 9–10, 29, 36, 82, 98, 147–9, 183–4, 188, 190, 193, 196, 203

Laplace, P. S. 2, 14–23, 25, 56, 85, 102, 147, 150–4, 165, 170, 184, 189–90, 196
Laurent, P. A. 140, 170
Legendre, A. M. 13, 22, 24–5, 30–2, 51–2, 54–6
Leibniz, G. W. 11
Libri, G. 105, 129, 132
Lindelöf, E. 4
Liouville, J. 138, 145
Lombardi, A. 173

Mittag-Leffler, G. 194

Noether, M. 2, 4

Osgood, W. F. 2, 4
Ostrogradskiĭ, M. V. 37, 86, 97, 111, 115, 193–5, 204

Paoli, P. 147, 151, 165, 170, 196
Parseval, M. A. 72–3, 105, 129
Piola, G. 157–8, 160, 169, 188
Poisson, S. D. 2, 14, 17–23, 25, 37–8, 43, 71, 75, 86–7, 105, 129–30, 189–90, 192
Pringsheim, A. 133

Riemann, G. F. B. 82, 188
Rouché, E. 172–3, 181, 185, 197–8
Ruffini, P. 80, 196

Smithies, F. 37, 50, 106
Stäckel, P. 4, 85

Terracini, A. 4
Timchenko, I. 4, 5
Truesdell, C. A. 4, 9

Valson, C. A. 4
Vernier, H. V. 105

Weierstrass, K. 144, 194

Yushkevich, A. P. 37, 97

Subject index